Molecular Biology Biochemistry and Biophysics
23

Editors:

*A. Kleinzeller, Philadelphia · G. F. Springer, Evanston
H. G. Wittmann, Berlin*

Advisory Editors:

*C. R. Cantor, New York · F. Cramer, Göttingen · F. Egami, Tokyo
M. Eigen, Göttingen · F. Gros, Paris · H. Gutfreund, Bristol
B. Hess, Dortmund · H. Jahrmärker, Munich · R. W. Jeanloz, Boston
E. Katzir, Rehovot · B. Keil, Gif-sur-Yvette · M. Klingenberg, Munich
I. M. Klotz, Evanston · F. Lynen, Martinsried/Munich
W. T. J. Morgan, London · K. Mühlethaler, Zurich · S. Ochoa, New York
G. Palmer, Houston · I. Pecht, Rehovot · R. R. Porter, Oxford
W. Reichardt, Tübingen · H. Tuppy, Vienna
J. Waldenström, Malmö*

M. Luckner · L. Nover · H. Böhm

Secondary Metabolism and Cell Differentiation

With 52 Figures

QH634.5
L8
1977

Springer-Verlag Berlin · Heidelberg · New York 1977

Professor Dr. Martin Luckner, Martin-Luther-Universität Halle-Wittenberg, Sektion Pharmazie, Weinbergweg, DDR 402 Halle/Saale

Dr. Lutz Nover, Akademie der Wissenschaften der DDR, Forschungszentrum für Molekularbiologie und Medizin, Institut für Biochemie der Pflanzen, Weinberg, DDR 402 Halle/Saale

Dr. Hartmut Böhm, Akademie der Wissenschaften der DDR, Forschungszentrum für Molekularbiologie und Medizin, Institut für Biochemie der Pflanzen, Weinberg, DDR 402 Halle/Saale

ISBN 3-540-08081-3 Springer-Verlag Berlin Heidelberg New York
ISBN 0-387-08081-3 Springer-Verlag New York Heidelberg Berlin

Library of Congress Cataloging in Publication Data. Luckner, Martin. Secondary metabolism and cell differentiation. (Molecular biology, biochemistry and biophysics; v. 23). Includes bibliographies and index. 1. Cell metabolism. 2. Cell differentiation. I. Nover, Lutz, 1941—, joint author. II. Böhm, Hartmut, 1936—, joint author. III. Title. IV. Series. [DNLM: 1. Cell differentiation. W1 Mo195T no. 23 19/QH607 L941s]. QH634.5.L8. 574.8′761. 76-53021.

This work is subject to copyright. All rights are reserved, whether the whole or part of the material is concerned, specifically those of translation, reprinting, re-use of illustrations, broadcasting, reproduction by photocopying machine or similar means, and storage in data banks. Under § 54 of the German Copyright Law, where copies are made for other than private use, a fee is payable to the publisher, the amount of the fee to be determined by agreement with the publisher.

© by Springer-Verlag Berlin · Heidelberg 1977.

The use of registered names, trademarks, etc. in this publication does not imply, even in the absence of a specific statement, that such names are exempt from the relevant protective laws and regulations and therefore free for general use.

Offsetprinting and bookbinding: Brühlsche Universitätsdruckerei, Gießen. 2131/3130-543210

Contents

I. Expression of Secondary Metabolism. An Aspect of Cell Specialization of Microorganisms, Higher Plants, and Animals. MARTIN LUCKNER and LUTZ NOVER 1

   A. Introduction 3
      1. Secondary Metabolism and Differentiation 3
      2. Investigation of the Molecular Biology of Differentiation 6
   B. Coordinate and Noncoordinate Formation of Enzymes of Secondary Metabolism 10
   C. Regulatory Effectors of Secondary Metabolism 16
      1. The Influence of Tryptophan and Tryptophan Analogs on the Expression of Ergolin Alkaloid Biosynthesis in *Claviceps* Species 17
      2. Stimulation of Streptomycin Formation in *Actinomyces streptomycini* by the A-Factor 22
      3. The Effect of Cinnamic Acids on Anthocyanin Synthesis in *Petunia hybrida* 24
      4. The Influence of the Light-Phytochrome System on the Regulation of Phenylpropanoid Metabolism in Higher Plants 26
      5. The Action of Adrenocorticotrophic Hormone on Corticosteroid Formation in the Adrenal Cortex ... 33
   D. Phase Dependence of Secondary Metabolism and the Organization of Differentiation Programs 37
      1. Integration of Alkaloid Metabolism into the Developmental Program of *Penicillium cyclopium* 39
          a) Growth, Conidia, and Alkaloid Formation 39
          b) Formation of the Enzymes of Alkaloid Metabolism during Conidiation 44
          c) Influence of Inhibitors of Gene Expression on Hyphal Cyclopenin-Cyclopenol Formation 46
          d) Glucose as Repressor of Idiophase Development . 53
          e) The Developmental Program of *Penicillium cyclopium* 55
      2. Biosynthesis of Secondary Products during Bacterial Sporulation 58
      3. Sequential Gene Expression in Secondary Metabolism of *Penicillium urticae*? 63
      4. Sequential Formation of Secondary Products during the Generative Development of Mucoraceous Fungi .. 65
      5. Expression of Secondary Metabolism in Developing Chloroplasts 70
      6. Secondary Product Formation during Microsporogenesis in Higher Plants 72
      7. Induction of Urea Cycle Enzymes as Part of the Thyroid Hormone-Stimulated Metamorphosis of *Rana catesbeiana* Tadpoles 74

8. The Formation of Tanning Agents during Ecdysone-
      Controlled Pupation of *Calliphora* Larvae 77
   E. Conclusions 79
   References 82

II. Secondary Metabolism in Cell Cultures of Higher Plants
    and Problems of Differentiation. HARTMUT BÖHM 104

   A. Introduction 105
   B. The Fate of Secondary Metabolism during Initiation
      of Plant Cell Cultures 106
   C. Realization of Secondary Metabolism in Plant Cell
      Cultures 107
      1. Triggering Factors 107
      2. Enzyme Activities 110
      3. Comparison with the Related Intact Plant 113
   D. Correlation between Secondary Metabolism and Cellular
      Structures 115
   E. Growth of Plant Cell Cultures and Formation of Sec-
      ondary Substances 118
   F. Concluding Remarks 119
   References 120

Subject Index 125

Abbreviations

ACTH: adrenocorticotrophic hormone
cAMP: cyclic adenosine-3',5'-monophosphate
CH: cycloheximide
CD: cyclopeptine dehydrogenase
DE: dehydrocyclopeptine epoxidase
6-MSA: 6-methyl salicylic acid
PAL: phenylalanine ammonia-lyase
P_i: inorganic phosphate
p.i.:   post inoculation
SCP: steroid carrier protein
TA: trisporic acids

I. Expression of Secondary Metabolism

An Aspect of Cell Specialization of Microorganisms, Higher Plants, and Animals

MARTIN LUCKNER and LUTZ NOVER

A. Introduction

1. Secondary Metabolism and Differentiation

In addition to the primary metabolic reactions, which are similar in all living beings (formation and breakdown of nucleic acids and proteins as well as of their precursors, of most carbohydrates, of some carboxylic acids, etc.), a vast number of metabolic pathways lead to the formation of compounds peculiar to a few species or even to a single chemical race only. These reactions, in accord with CZAPEK (1921) and PAECH (1950), are summed up under the term "secondary metabolism", and their products are called "secondary metabolites."

The wide variety of secondary products formed in nature includes such well-known groups as alkaloids, antibiotics, cardiac glycosides, tannins, saponins, volatile oils, and others. A considerable number of them are of economic importance in therapeutics or technology. Although secondary products are produced by microorganisms, higher plants, and animals (cf. LUCKNER, 1972), most of the substances are found in the plant kingdom. The lack of mechanisms for true excretion in higher plants may result in this unequal distribution, the "waste products" of metabolism in plants instead being accumulated in the vacuoles, the cell walls, or in special excretory cells or spaces of the organism ("metabolic excretion," cf. FREY-WYSSLING, 1935, 1970; MOTHES, 1966a, b, 1972; LUCKNER et al., 1976.

Many secondary substances have, however, a direct biologic function. They can be regulatory effectors, e.g., the hormones of plants and animals (LUCKNER, 1971) or the hormonelike substances in bacteria (cf. Chap. C2), or they can be ecologically significant substances (pigments and pheromones, substances involved in defense mechanisms against bacteria, pathogenic fungi, and other enemies, factors enabling life in special ecologic niches; cf. SONDHEIMER and SIMEONE, 1970; LUCKNER et al., 1976).

The biosynthesis of secondary products is usually restricted to specific developmental stages of the organism and the specialized cells, respectively. This phenomenon was recently shown, in several instances, to be due to the phase-dependent formation of the corresponding enzymes, i.e., expression of secondary metabolism is based on a differentiation process.

The proteins formed as a result of the differentiation processes can be classified according to their biologic significance for and function in the producing cell, as seen in Table 1. According to this classification, secondary metabolism may be defined as *"biosynthesis, transformation and degradation of endogenous compounds by specialization proteins."*

The following is a brief outline of a general concept of differentiation that will be the basis for the discussion about the regulatory aspects of secondary metabolism.

In order to focus attention upon the similarities of the basic mechanisms, the term "differentiation" encompasses all processes by which cells become different from each other through unequal expression of the same genetic material, i.e., all kinds of differential gene expression, irrespective of the cell type involved (prokaryotic or eukaryotic, embryonic or specialized cells of unicellular or multicellular organisms) and irrespective of the lifetime of the differentiation product.

Differentiation in the preceding sense has five main aspects:

1. The triggering of differential gene expression by extracellular or intracellular factors; processes of signal reception and transformation
2. The mechanism of differential gene expression
3. The differential breakdown of gene expression products
4. The coordination of individual steps of differential gene expression to differentiation programs
5. The coordination by intercellular interactions of the processes within individual cells (differentiation) to the developmental programs of tissues, organs, and organisms.

Gene expression includes the following partial processes:

1. Gene activation and transcription of DNA into messenger RNA (mRNA) and its precursors, respectively
2. Processing of mRNA precursors and transport of its products to the sites of protein synthesis
3. Messenger RNA translation
4. Activation of biologically inactive proteinogens by enzymatic modification (proteinogen processing).

All partial processes offer multiple sites for specific positive or negative control by regulatory proteins and/or RNAs that are able to interact with different components of the gene expression chain. The activity of these proteins frequently depends on interaction with effectors, which act as a kind of signal and carry information from inside or outside the cell to the sites where gene expression takes place.

Two types of effectors may be distinguished (cf. Chap. C):
(1) substrate-like effectors and (2) nonsubstrate-like effectors.

Substrate-like effectors have a direct relation to the biologic function of the proteins whose synthesis they act upon. They include substrates or products of enzymes or enzymatic chains, prosthetic groups of proteins, etc. Substrate-like effectors generally influence the synthesis of a restricted number of proteins that have the same biologic functions in very different organisms, though their detailed modes of action may differ from one organism to the other.

Nonsubstrate-like effectors, in the sense indicated above, have no direct relationship to the biologic functions of the proteins whose synthesis they control. They include the plant and animal hormones, cyclic adenosine-3'5'-monophosphate (cAMP) and similar

Table 1. Differentiation products

A. Primary metabolic proteins	B. Specialization proteins
Proteins which themselves or the products of which are typical for the cells of all organisms or at least large groups of organisms and which are of direct importance for the existence and reproduction of the cell	Proteins that are found in some cells only and that themselves or the products of which are frequently without direct importance for the cell producing them, though they may have a vital function for the total organism
1. Enzymes	1. Enzymes
Enzymes of primary metabolism	Enzymes that enable the cell to perform special functions, e.g., *enzymes of secondary metabolism*, enzymes with a function in morphogenesis, exoenzymes excreted by special cells of almost all organisms, special catabolic enzymes that degrade drugs, special nutrients, etc.
2. Nonenzymatic proteins	2. Nonenzymatic proteins
Structural proteins of the cell membrane and the protoplasm, ribosomal structural proteins, regulatory proteins, e.g., histones, acidic proteins of chromatin, etc.	Proteins that enable the cell to perform special functions within the organism, e.g., hemoglobin, muscle proteins, storage proteins, cilial proteins and exoproteins such as collagen, blood plasma proteins, antibodies, milk proteins, and some peptide hormones, structural and regulatory proteins required for the specialized function of the cell

compounds. The particular effect of these substances is often restricted to closely related groups of organisms and, within one organism, to a limited number of cells, the so-called target cells. Depending on the stage of differentiation, cells of different tissues may respond to the same nonsubstrate-like effector with totally different changes of the protein synthesis pattern.

Connecting the individual steps of differential gene expression to relatively complicated differentiation programs (cf. Chap. D) is an essential aspect of differentiation. The individual steps are joined by regulatory effectors, regulatory proteins, or RNAs. Tightly linked with the realization of differentiation programs are the phenomena of (1) competence, i.e., the stage-restricted capability of cells to respond to the particular signals by expression of distinct parts of their genetic material and of (2) determination, i.e., the commitment of cells to certain differentiation processes or programs by prior changes of their chemical composition.

Within the frame just outlined, the investigation of secondary metabolism as an aspect of differentiation requires a profound knowledge of the chemistry and biochemistry of the compounds under consideration, including possible intermediates of the biosynthetic chains. In addition, at least some of the enzymes involved should be measurable in vitro and sufficiently characterized. When, however, these requirements are fulfilled, experimental work with secondary products frequently offers the opportunity to simply and reliably quantify the differentiation characteristics and to effect manifold metabolic and mutational interferences with the process without serious influence on the viability of the cell.

2. Investigation of the Molecular Biology of Differentiation

In recent years, a number of potent in vitro methods have markedly improved our knowledge of the molecular biology of differential gene expression and hence of the basis of differentiation generally. These methods are listed below together with a few selected references:

1. Isolation or in vitro synthesis of structural genes or regulatory sequences (FOURNIER et al., 1973; WILCOX et al., 1974; FORGET et al., 1975; LIS and SCHLEIF, 1975; MAJORS, 1975; MEYER et al., 1975)
2. Sequencing of regulatory parts of DNA and RNA (GILBERT and MAXAM, 1973; BERTRAND et al., 1975; DICKSON et al., 1975; MEYER et al., 1975; PIRROTTA, 1975; WALZ and PIRROTTA, 1975; SZYBALSKI, 1977)
3. Purification and chemical as well as functional characterization of regulatory proteins (VON HIPPEL and McGHEE, 1972; HUANG, 1972; GROS, 1974; O'MALLEY and MEANS, 1974; TYLER et al., 1974; FAILLA et al., 1975)
4. Direct electronmicroscopic visualization of active genes (BEERMANN, 1972; HAMKALO and MILLER, 1973)
5. Hybridization of RNA with DNA or chromatin, for the quantification of mRNA or demonstration of differential gene activity (SZYBALSKI et al., 1970; KIM, 1972; STEFFENSEN and WIMBER, 1972; IMAMOTO, 1973; WILCOX et al., 1974; FORGET et al., 1975; PAUL and GILMOUR; TSAI and O'MALLEY, 1977; HAHLBROCK, 1977)
6. Detection and quantification of polysomes primed by selected mRNAs (SARKAR and MOSCONA, 1973; BRAWERMAN, 1974; PALADE, 1975; SHAPIRO and SCHIMKE, 1975)
7. In vitro transcription of selected genes (IMAMOTO, 1973; GROS, 1974; FORGET et al., 1975; MAJORS, 1975; MEYER et al., 1975; TSAI and O'MALLEY, 1977)
8. Isolation and in vitro translation of mRNAs (BRAWERMAN, 1974; HUEZ et al., 1974; PALMITER and CAREY, 1974; BOIME et al., 1975; HARTLEY et al., 1975; POYTON and GROOT, 1975; SHAPIRO and SCHIMKE, 1975; VERMA et al., 1975; WOO and O'MALLEY, 1975)
9. Composition of coupled in vitro transcription-translation systems (GROS, 1974; WILCOX et al., 1974; BREINDL and HOLLAND, 1975; KUNG et al., 1975).

Thus far, these in vitro methods have failed to reflect essential features of the overall process of differentiation, which are bound at least partly to the structural organization of the cell, e.g., processes of signal reception and transformation, organization of differentiation processes, and the ordered and regulated procedure of transcription, RNA processing and translation, and possibly proteinogen processing in eukaryotic systems (cf. the model of cascade regulation of gene expression by SCHERRER, 1973).

Besides these in vitro methods, there are others aimed at measuring the increase in the cell of the final products of differential gene expression, i.e., of the biologically active proteins, by direct cytophotometric methods (HOLTZER et al., 1973) or by in vitro determination of the enzyme activity. However, only isotope-labeling (HU et al., 1962; FILNER et al., 1969; HAHLBROCK and SCHRÖDER, 1975; RICKWOOD and BIRNIE, 1975; WELLMANN and SCHOPFER, 1975) and/or immunologic experiments (YALOW and BERSON, 1964; MOSCONA et al., 1972; JOH et al., 1973; FRAGOULIS and SEKERIS, 1975a) have shown unequivocally the de novo formation of a protein, since the activity increase measured in vitro may be due to other processes, e.g., proteinogen processing, cessation of degradation of the protein, etc.

An additional, frequently applied procedure for investigating differentiation processes is the use, in vitro and in vivo, of inhibitors of gene expression, usually in combination with the methods discussed previously. Use of the inhibitors offers the opportunity to obtain data on the regulatory details and dynamics of differentiation processes with moderate experimental expenses. However, a prerequisite for this type of experiment is sufficient knowledge about the qualitative and quantitative data of the inhibitor effects on the organism under investigation, in order to distinguish between primary effects on gene expression and secondary effects frequently evoked because of interference with other cellular processes. On the other hand, "peculiar effects" of the inhibitors may indicate special regulatory features of the process under investigation. Thus there are a number of examples where inhibitors evoke stimulatory rather than inhibitory effects (cf. the summary of TOMKINS et al., 1972 as well as LIGHT, 1970; JONES and WEISSBACH, 1970; MANCINELLI et al., 1972; KILLEWICH et al., 1975). Examples of secondary metabolism are discussed in Chapts. C4 and D1c.

Table 2 is a summary of the experimental results demonstrating that the expression of secondary metabolism as part of cell spezialization is bound to the de novo synthesis of RNA and protein. With two exceptions (cf. pp. 30 and 79) most results were obtained by in vivo and in vitro measurements of enzyme activities and/or application of inhibitors of gene expression only. This situation reflects the very poor experimental state of the molecular biology of secondary metabolism as compared to that of other differentiation processes.

Table 2. Investigation of differential gene expression in secondary metabolism

Organism or system	Secondary products whose synthesis has been investigated	Methods applied[a]	References
Bacteria			
Bacillus spec.	Peptide antibiotics	A B	KURAHASHI, 1974 WEINBERG and TONNIS, 1966, 1967; KURAHASHI, 1974 (cf. Chap. D2)
Bacillus megaterium	Dipicolinic acid	B	MANDELSTAM, 1976 (cf. Chap. D2)
Bacillus subtilis	Sulfolactic acid	B	WOOD, 1971 (cf. Chap. D2)
Streptomyces antibioticus	Actinomycin chromophore	B	MARSHALL et al., 1968
Streptomyces hydrogenans	Steroids	B	BETZ and TRÄGER, 1975
Fungi			
Penicillium cyclopium	Benzodiazepine alkaloids	A B	LUCKNER, 1977 NOVER and MÜLLER, 1975; EL KOUSY et al., 1975; NOVER and LUCKNER, 1976 (cf. Chap. D1)
Penicillium urticae	6-Methylsalicylic acid, patulin	A B	LIGHT, 1967, 1970; BU'LOCK et al., 1969 (cf. Chap. D3)
Aspergillus fumigatus	Orsellinic acid and other polyketides	B	PACKTER and COLLINS, 1974; WARD and PACKTER, 1974
Blakeslea trispora	Carotenoids, trisporic acids	B	FEOFILOWA and BEKHTEREVA, 1976 (cf. Chap. D4)
Claviceps spec.	Ergoline alkaloids	A	HEINSTEIN et al., 1971; ERGE et al., 1973 (cf. Chap. C1)
Higher Plants			
Various higher plants and corresponding cell cultures	Cinnamic acid and flavonoid derivatives	A B C D	GRISEBACH and HAHLBROCK, 1974; McCLURE, 1975; HAHLBROCK and RAGG, 1975; HAHLBROCK and SCHRÖDER, 1975 (cf. Chap. C4) HAHLBROCK, 1977

Table 2. (Continued)

Organism or system	Secondary products whose synthesis has been investigated	Methods applied[a]	References
Phaseolus vulgaris	Phaseollin	B	HESS and HADWIGER, 1971
Amaranthus caudatus, Celosia plumosa	Betalains	B	GIUDICI DE NICOLA et al., 1973; KÖHLER, 1975
Pisum sativum, Malus silvestris	Ethylene	B	LIEBERMANN and KUNISHI, 1975
Animals			
Mammals	Corticosteroids	B	SCHULSTER, 1974 (cf. Chap. C5)
Rana catesbeiana	Urea	A	COHEN and BROWN, 1960 (cf. Chap. D7)
Calliphora erythrocephala	Tanning agents in the pupal cuticula	A	KARLSON and SCHWEIGER, 1961; FRAGOULIS and SEKERIS, 1975a;
		B	SEKERIS and KARLSON, 1964; SHAAYA and SEKERIS, 1971;
		D	FRAGOULIS and SEKERIS, 1975b;
		E	FRAGOULIS and SEKERIS, 1975a (cf. Chap. D8)
Neoblastoma tumor cells	Acetylcholine, dopa	A	HAFFKE and SEEDS, 1975

[a] A: Determination of in vitro activity of corresponding enzymes; comparison of rates of secondary product formation in vivo with in vitro enzyme activities.
B: Measurement of influence of inhibitors of gene expression on formation of secondary products or of enzymes involved in their biosynthesis.
C: Isotope labeling of enzymes catalyzing secondary product formation to demonstrate their de novo formation.
D: Translation of mRNA in vitro to show phase-dependent presence of mRNA species coding secondary metabolic enzymes.
E: Immunologic measurements of enzymes of secondary metabolism to demonstrate their de novo formation at specific developmental stages.

B. Coordinate and Noncoordinate Formation of Enzymes of Secondary Metabolism

The biochemical balance of living cells requires exact, mutual harmony among all the different parts of the overall metabolic process. Constant improvement of this regulatory interdependence has been an essential aspect of the evolution of primary metabolism. Besides multiple metabolic controls by allosteric regulation of enzyme activities, the corresponding proteins generally are produced in a coordinated manner, leading to constant proportions of the individual enzymes participating in a common biosynthetic pathway.

Genetic and biochemical experiments in bacteria have shown that the genetic material of such pathways is grouped into regulatory units, parts of which may be localized at different sites on the genome. The clustering of genes in operon-like structures is not a prerequisite for the coordinated regulation of enzyme synthesis. In fact, operons comprising all enzymes of a given metabolic pathway are the exception rather than the rule. The regulatory interdependence between nonclustered genes in the regulons may be very tight (cf. the regulation of arabinose metabolism in *Escherichia coli*, ENGELSBERG and WILCOX, 1974) or it may be restricted to specific metabolic situations (KANE et al., 1972). A special aspect of the latter phenomenon is the coordinated regulatory influence of some group-specific effectors on gene expression in parts of the catabolic and anabolic processes of bacteria. Such are cAMP, the effector of catabolite repression (RICKENBERG, 1974), guanosine-3'-diphosphate-5'-diphosphate, the effector of the stringent-relaxed control of amino acid biosynthesis (STEPHENS et al., 1975), and the effector of the so-called nitrogen catabolite repression (FOOR et al., 1975).

Though the genetic and detailed biochemical evidence for the coordinated expression of gene groups in eukaryotic cells is scarce and circumstantial (CARSIOTIS et al., 1970; GREENGARD, 1971; GUERZONI, 1972; METZENBERG, 1972), it can be anticipated that the extreme complexity of the eukaryotic organism requires the flexible integration of genes into groups with coordinate expression depending on the differentiation stage of the cell. This means a considerably increasing complexity of the regulatory interactions between different parts of the genetic material, a phenomenon that can be illustrated by the coordinated regulation of enzymes and enzyme systems participating in fatty acid biosynthesis in rat liver (LANE and MOSS, 1971).

The following contains examples that illustrate the extent of coordinated enzyme formation in secondary metabolism as elaborated by studying the dynamics of enzyme activities during developmental stages. It should be noted that the problem will be discussed further in several sections of Chap. D, which deal with the expression of secondary metabolism as part of differentiation programs.

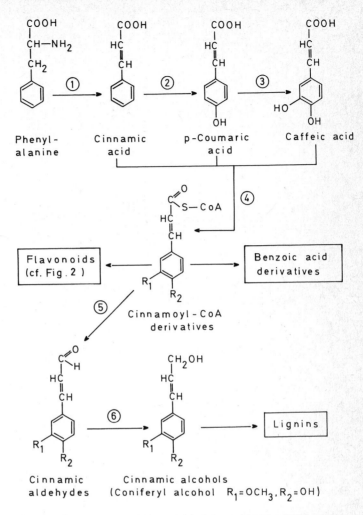

Fig. 1. Cinnamic acid metabolism (from GRISEBACH and HAHLBROCK, 1974; STAFFORD, 1974). (1) Phenylalanine ammonia-lyase (PAL), (2) cinnamate 4-hydroxylase, (3) 4-hydroxycinnamate 3-hydroxylase; phenolase, (4) cinnamate (p-coumarate): CoA ligase, (5) cinnamoyl-CoA reductase, (6) aromatic aldehyde reductase; coniferyl alcohol dehydrogenase

In vitro characterization and measurement of many enzymes involved in flavonoid metabolism (Figs. 1 and 2) have shown that, in higher plants as well as in plant cell cultures, groups of enzymes obviously show coordinated regulation. In parsley cell cultures, in which the biosynthesis of flavonoids, e.g., apiin and graveobiosid B, is inducible by irradiation (HAHLBROCK and WELLMANN, 1970; WELLMANN, 1974, 1975), the dynamics of the enzymatic activities reveal two groups: Group I comprises the enzymes of cinnamic acid metabolism (Fig. 1, Nos. 1-3), group II, those of the special pathway leading from cinnamoyl-CoA derivatives to the flavonoids themselves (Fig. 2, Nos. 2, 4, 5, 6, and a flavonoid specific methyltransferase; HAHLBROCK et al., 1976; for a detailed dis-

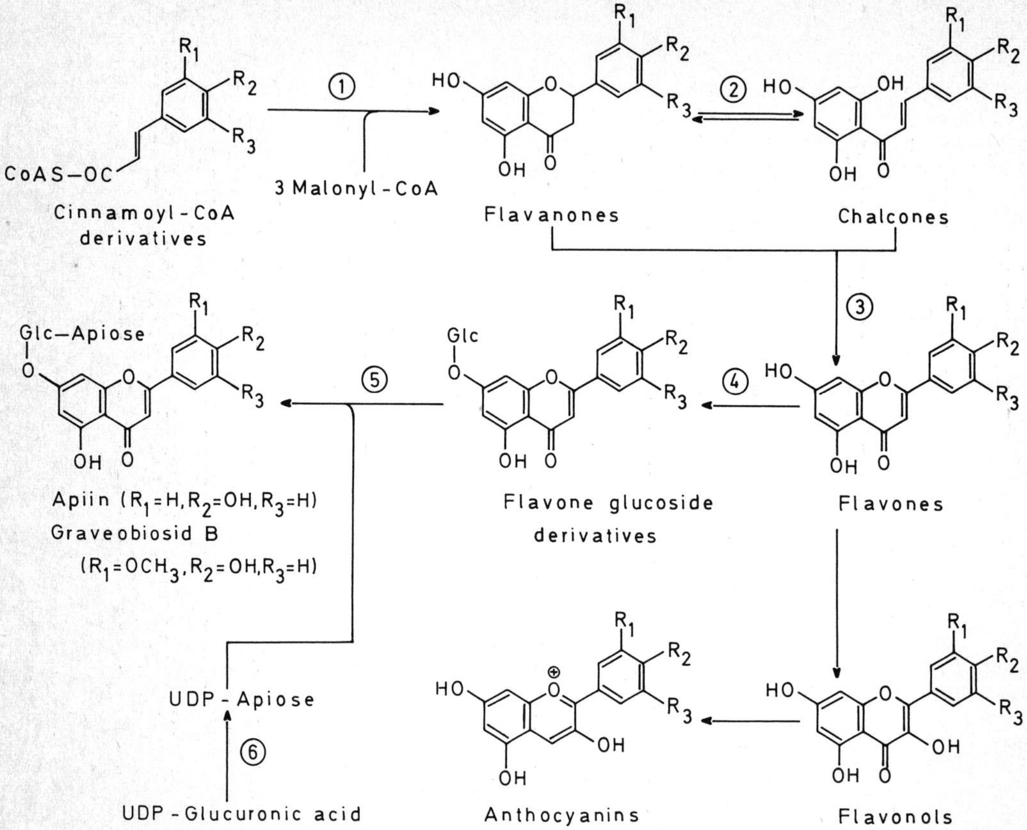

Fig. 2. Formation of flavonoids from cinnamoyl derivatives (from GRISEBACH and HAHLBROCK, 1974). Enzymes involved: (1) Flavanone synthetase, (2) chalcone-flavanone isomerase, (3) chalcone-flavanone oxidase, (4) glucosyltransferase, (5) apiosyltransferase, (6) UDP-apiose synthetase.

Further cinnamoyl derivatives and flavonoids treated with in the following chapters

Groups	R_1 H, R_2 H, R_3 H	R_1 H, R_2 OH, R_3 H	R_1 OH, R_2 OH, R_3 H	R_1 OCH$_3$, R_2 OH, R_3 H
Cinnamoyl-CoA derivatives	Cinnamoyl-CoA	p-Coumaroyl-CoA	Caffeoyl-CoA	Feruloyl-CoA
Chalcones		2',4,4',6'-Tetrahydroxy-chalcone	2',3,4,4',6'-Pentahydroxy-chalcone	
Flavones		Apigenin		
Flavonols		Kaempferol	Quercetin	Isorhamnetin
Anthocyanins			Cyanidin	Peonidin

cussion cf. BÖHM, Chap. C2). As many as three types of activity profiles for enzymes involved in phenylpropanoid metabolism were detected in petunia cell cultures (Fig. 3). Soon after inoculation, there is a coordinated increase of group I enzymes (Fig. 1, Nos. 1-4); the activity of chalcone-flavanone isomerase, a representative of group II, peaks about the 4th day, and the activity of coniferyl alcohol dehydrogenase, an enzyme involved in lignin biosynthesis (Fig. 1, No. 6), increases between the 10th and 17th days of culture development.

Some reports, however, suggest that the coordination pattern found in cell cultures may not be applicable to intact plants. In cotyledons and leaves of parsley seedlings, phenylalanine ammonia-lyase (PAL) and several group II enzymes show a regulatory behavior that indicates interdependence (cf. Fig. 4). Furthermore, in buckwheat, red cabbage, parsley, radish, and white mustard seedlings, PAL is induced by light, whereas p-coumaric acid CoA: ligase, another group I enzyme, is not (McCLURE and GROSS, 1975).

An interesting example, which demonstrates that regulatory behavior changes during the course of development, is found in *Papaver bracteatum* (BÖHM, 1971). In a branched pathway (Fig. 5), two final products of alkaloid metabolism, thebaine and alpinigenine, are formed in this plant. Three genetic varieties with differing capabilities for alpinigenine synthesis were selected from a large population of *P. bracteatum* plants. In the e^+-type, thebaine and alpinigenine are formed during the entire lifetime, whereas no alpinigenine is synthesized in the e^--type. Plants of a third variety (e^h-type) produce alpinigenine only in their youth.

Cross experiments revealed that the e^+-type is dominant over the e^h- and the e^--types and that the e^h-type is dominant over the

---

R_1	R_2	R_3	R_1	R_2	R_3	R_1	R_2	R_3
OH	OH	OH	OCH_3	OH	OH	OCH_3	OH	OCH_3

Trihydroxy-cinnamoyl CoA	5-Hydroxy-feruloyl -CoA	Sinapyl-CoA

Delphinidin   Petunidin   Malvidin

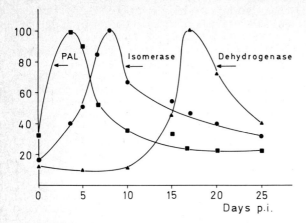

Fig. 3. Activities of enzymes of phenylpropanoid metabolism in petunia cell cultures (from RANJEVA et al., 1975). Cells were grown in medium containing ^{6}N-benzyladenine (0.2 mg/l). ■ PAL as representative of group I enzymes (100 = 3 mU/mg protein); ● Chalcone-flavanone isomerase as representative of group II enzymes (100 = 3 mU/mg protein); ▲ Coniferyl alcohol dehydrogenase as representative of enzymes catalyzing lignification (100 = 2.5 mU/mg protein)

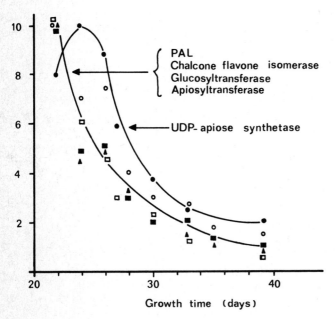

Fig. 4. Enzyme activities in developing parsley leaves (from HAHLBROCK et al., 1971b, redrawn). Activities are calculated in enzyme units/g fresh weight. Phenylalanine ammonia-lyase (PAL) (10 = 0.04 U) (o), chalcone flavone isomerase (10 = 3.8 U) (■), UDP-glucose: apigenin 7-O-glucosyltransferase (10 = 0.25 U) (▲), and UDP-apiose: apigenin glucoside apiosyltransferase (10 = 2.8 · 10^{-5} U) (□) form a regulatory group. UDP-apiose synthetase (10 = 0.04 U) (●) activity seems to show independent regulation

e^--type. The e^h-characteristic shows monohybrid inheritance. Crosses between several e^h-types never lead to restitution of the e^+-phenotype. By feeding intermediates of the alpinigenine branch of alkaloid metabolism to old plants of the e^h-type, it was found that (+)-reticuline, though incorporated into thebaine, was not converted to alpinigenine. In contrast, tetrahydropalmatine or protopine derivatives, which are intermediates between (+)-reticuline and alpinigenine, give rise to the latter compound. No conversion of the intermediates was found in e^--plants of different ages, whereas all compounds fed to plants of the e^+-type were incorporated into alpinigenine.

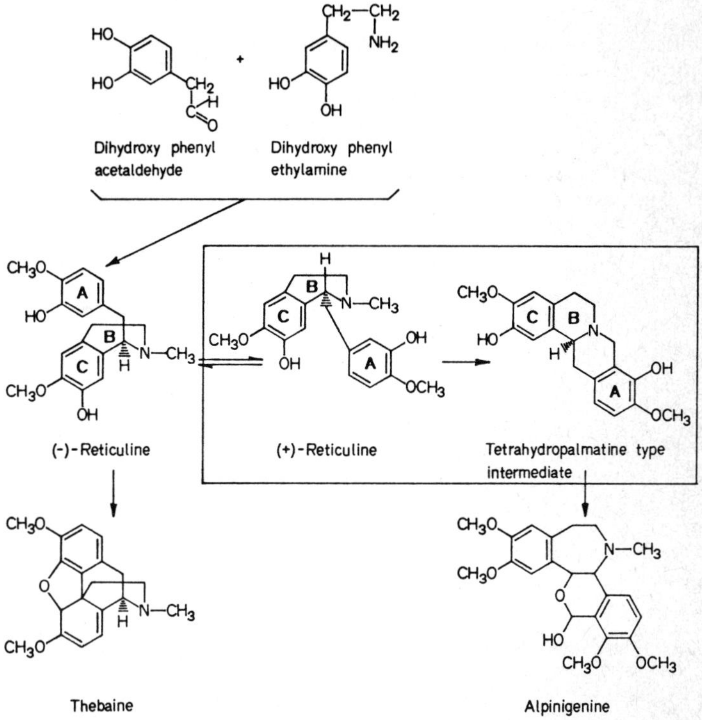

Fig. 5. Pathway leading to alkaloids thebaine and alpinigenine in *Papaver bracteatum* (from BÖHM, 1971, redrawn). Part of biosynthetic chain with key function in regulation of alpinigenine formation lies within square

Though these results are far from sufficient for a final evaluation of the peculiarities of the e^h-type, the following interpretation appears reasonable (BÖHM, unpublished data). The genetic difference distinguishing the e^h-type from the e^+-type very likely involves only one locus. Because alpinigenine biosynthesis proceeds normally in homozygous young plants of the e^h-type as well as in e^+ x e^h hybrids over the entire period of life, all enzymes of the pathway are evidently intact in young e^h plants; hence, mechanisms for selective degradation of alpinigenine or for an enzyme involved in the (+)-reticuline-tetrahydropalmatine conversion are excluded. The results indicate, rather, that the mutation in the e^h-type involves the formation or nonformation of a regulatory molecule (effector, regulatory protein), that influences the transformation of (+)-reticuline into the tetrahydropalmatine type intermediate, i.e., it would seem that, in the plants of the e^h-type, an important group of enzymes of the alpinigenine branch is regulated independently from the remaining enzymes of alkaloid metabolism.

C. Regulatory Effectors of Secondary Metabolism

A large number of chemical and physical signals bring about the profitable interaction of cells with their surrounding, i.e., the adaptation of cells to the metabolic and functional requirements imposed by the nutrient milieu or by the particular role within the multicellular organism. Many of these signals act directly or indirectly as effectors of differential gene expression. They may be classified as nonsubstrate-like and substrate-like (cf. Chap. A).

Examples of the action of *substrate-like* effectors are the stimulation of synthesis of β-galactosidase in bacteria (RICHMOND, 1968; JOBE and BOURGEOIS, 1972; BECKWITH and ROSSOW, 1974) and ascomycetes (ININGER and NOVER, 1975) by β-galactosides, of nitrate reductase in bacteria (SHOWE and DE MOSS, 1968), fungi (NASON and EVANS, 1953; COVE and PATEMAN, 1969), and higher plants (FILNER et al., 1969) by nitrate, and of aryl hydroxylase in mammalian liver by phenobarbital and polycyclic hydrocarbons (NEBERT et al., 1972; SCHIMKE, 1973). In amino acid biosynthesis in bacteria and fungi (UMBARGER, 1969; VOGEL, 1971), or heme biosynthesis in bacteria and higher animals (GRANICK, 1967), the formation of the corresponding enzyme systems and key enzymes, respectively, is repressed by the end products (amino acids, heme).

The second group, the *nonsubstrate-like* effectors, is best represented by some animal hormones (TOMKINS et al., 1969; GREENGARD, 1971; MOSCONA, 1973; O'MALLEY and MEANS, 1974; LISSITZKY et al., 1975; GUIDOTTI et al., 1975), by some plant hormones, e.g., gibberellic acid (GALSKY and LIPINCOTT, 1969; TUAN-HUA HO and VARNER, 1974), by cAMP as a mediator of animal hormone effects (LISSITZKY et al., 1975; GUIDOTTI et al., 1975) or as a trigger of slime mould development (ASHWORTH, 1971; BONNER, 1971; KILLICK and WRIGHT, 1974; DARMON et al., 1975), by the light-phytochrome system regulating a vast number of differentiation processes in plants (BRIGGS and RICE, 1972; MITRAKOS and SHROPSHIRE, 1972; MOHR and SITTE, 1972) and by guanosine-3'-diphosphate-5'-diphosphate, the glutamine/glutamic acid ratio and cAMP as the effectors of amino acid starvation as well as nitrogen and carbon catabolite repression in bacteria (RICKENBERG, 1974; CASHEL, 1975; STEPHENS et al., 1975; MAGASANIK, 1977).

This classification into substrate-like and nonsubstrate-like effectors is, however, not a strict one and above all does not imply a principal difference in mode of action. It is complicated by the following facts:

1. In most cases, the actual regulatory effector of a gene expression process is unknown. The signals present in the extracellular milieu either must be chemically transformed to yield the intracellular effector, as is the case with some substrate-like effectors in bacteria (BECKWITH and ROSSOW, 1974) or they are only the first link of a signal transformation chain causing increase or decrease of the cellular concentration of the actual effector. The latter situation is found with some animal hormones (LISSITZKY et al., 1975; GUIDOTTI et al., 1975), with signals effective in C- and N-catabolite repression in bacteria (RICKENBERG, 1974; ENGELSBERG and WILCOX, 1974; MAGASANIK, 1977) or with the action of light on the reversible transformation of phytochrome in plants (BRIGGS and RICE, 1972).
2. Frequently, the effectors influence a multiplicity of gene expression processes, particularly those effectors acting on differentiation programs (cf. Chap. D) or those involved in catabolite repression in bacteria (RICKENBERG, 1974; ENGELSBERG and WILCOX, 1974; MAGASANIK, 1977).

The regulatory action of effectors is intimately connected with problems of cellular compartmentation. It is reasonable to assume that a positive reaction following the extracellular addition of an effector can be obtained only if the cell is capable of responding (competence), if the level of the effector in the regulatory compartment is suboptimal, and if the effector is able to enter this compartment. There is experimental evidence demonstrating the essential role of effector compartmentation for some substrate-like effectors (SERCARZ and GORINI, 1964; MATCHETT and DEMOSS, 1964; FILNER, 1969).

Table 3 and the following chapter contain examples that show the stimulation of secondary metabolism by both substrate-like and nonsubstrate-like effectors. Though in most cases there is a rather superficial knowledge about the mode of action it is already apparent that the effectors do not act directly on the expression of secondary metabolism. Rather, they stimulate or trigger differentiation programs that include the formation of enzymes catalyzing the formation of the secondary products (cf. Chap. D).

1. The Influence of Tryptophan and Tryptophan Analogs on the Expression of Ergolin Alkaloid Biosynthesis in *Claviceps* Species

As early as 1964 it appeared that tryptophan may play a dual role in the biosynthesis of ergolin alkaloids in *Claviceps* species: Besides its function as an alkaloid precursor, it is also a regulatory effector stimulating the rate of alkaloid formation (FLOSS and MOTHES, 1964). This stimulating effect was also found for some tryptophan analogs, which are not incorporated into the alkaloids, such as C-methylated tryptophans, homotryptophan, bishomotryptophan, and thiotryptophan (cf. FLOSS et al., 1974; KRUPINSKI et al., 1976; ROBBERS and FLOSS, 1976). The efficiency of the stimulators was especially pronounced when using nutrient

Table 3. Effectors promoting the expression of secondary metabolism

Organism or system	Secondary products or enzymes whose synthesis is increased	Effector Name	Substrate-like	Nonsubstrate-like	References
Bacteria					
Streptomyces hydrogenans	20β-Hydroxysteroid dehydrogenase	11β,21-Dihydroxy-4,17(20)pregnadien-3-on	+		BETZ and TRÄGER, 1975
Actinomyces streptomycini	Streptomycin	A-factor		+	KHOKHLOV et al., 1973, 1976 (cf. Chap. C2)
Fungi					
Aspergillus fumigatus, *Claviceps paspali*, *C. fusiformis*	Ergolin alkaloids	Tryptophan and analogs	+		RAO and PATEL, 1974; FLOSS et al., 1974; ROBBERS and FLOSS, 1976 (cf. Chap. C1)
Penicillium cyclopium	Benzodiazepine alkaloids, cyclopeptine dehydrogenase	Phenylalanine and analogs	+		DUNKEL et al., 1976 (cf. Chap. D1e)
		Mycelial factors		+	
Blakeslea trispora, *Rhodotorula* spec. and other carotinogenic microorganisms	Carotenoids	β-Jonone	+		FEOFILOWA and BEKHTEREVA, 1976
		2-(4-Chlorophenyl-thio)-triethyl-amine and other amines with RCH_2NEt_2 group		+	HAYMAN et al., 1974; POLING et al., 1975

Table 3 (Continued)

Mucorales (*Blakeslea* spec., *Phycomyces* spec., *Mucor* spec.)	β-Carotene and its oxygenated polymer sporopollenin	Trisporic acid	+	GOODAY, 1974 (cf. Chap. D4)

Higher plants

Different plant species and plant cell cultures	Cinnamic acid derivatives, flavonoids, lignin and enzymes involved in biosynthesis of these compounds	Auxins[a]	+	McCLURE, 1975; ALFERMANN and REINHARD, 1971, 1976
		Cytokinins[a]	+	PECKET and HATHOUT BASSIM, 1974; McCLURE, 1975; ZUCKER, 1972; SMITH, 1973; CAMM and TOWERS, 1973; McCLURE, 1975
		Phytochrome	+	(cf. Chap. C4)
Sorghum vulgare		Ethylene[a,b]	+	CRAKER and WETHERBEE, 1973; McCLURE, 1975; CRAKER, 1975
Daucus carota		Gibberellic acid[a]	+	SEITZ and HEINZMANN, 1975; McCLURE, 1975
Petunia hybrida		Cinnamic acid and derivatives	+	HESS, 1967a, b, 1968 (cf. Chap. C3)
Amaranthus tricolor, A. caudatus, Celosia plumosa	Betalains	Phytochrome	+	GIUDICI DE NICOLA et al., 1973, 1975; COLOMAS and BULARD, 1975
		Cytokinins[a]	+	PIATTELLI et al., 1971; GIUDICI DE NICOLA et al., 1973
		cAMP[c]	+	RAST et al., 1973
Different plant species	Ethylene	Phytochrome[a]	+	CRAKER et al., 1973; BURG, 1973
		Auxins	+	KANG et al., 1971; SAKAI and IMASEKI, 1971; BURG, 1973; LIEBERMANN and KUNISHI, 1975

Table 3. (Continued)

Organism or system	Secondary products or enzymes whose synthesis is increased	Effector Name	Sub-strate-like	Nonsub-strate-like	References
Morinda citrifolia cell cultures	Anthraquinones	Auxins		+	ZENK et al., 1975
Cucurbita moschata, Lycopersicon esculentum, Citrus paradisi	Carotenoids	2-(4-Chlorophe-nylthio)-triethyl-amine and other amines with RCH$_2$NEt$_2$ group		+	RABINOWITCH and RUDICH, 1972; POLING et al., 1975; SEYAMA and SPLITTSTOESSER, 1975
Animals					
Mammals	Corticosteroids	ACTH, cAMP		+	SCHULSTER, 1974 (cf. Chap. C5)
Neoblastoma tumor cells	Acetylcholine esterase	Acetylcholine	+		HAFFKE and SEEDS, 1975
	Choline acetyltrans-ferase, acetylcholine esterase, tyrosine hydroxylase	cAMP, papaverine		+	HAFFKE and SEEDS, 1975
Rana catesbeiana tadpoles	Enzymes of urea production	3,3',5'-Triiodo-thyronine		+	FRIEDEN and JUST, 1970; TATA, 1970 (cf. Chap. D7)
Calliphora erythrocephala	Dopadecarboxylase, phenoloxidase	β-ecdysone		+	KARLSON and SCHWEIGER, 1961; FRAGOULIS and SEKERIS, 1975a (cf. Chap. D8)

[a] Responses are variable and dose-dependent. Under certain conditions, formation of secondary products is inhibited. Footnotes b and c see page 21.

solutions that caused a low rate of alkaloid synthesis. For
instance, the low alkaloid production rate in media with high
phosphate content is partly overcome by addition of tryptophan
derivatives. For two of the enzymes involved in ergolin alkaloid
synthesis - dimethylallyl-pyrophosphate: tryptophan dimethylallyl
transferase (DMAT synthetase) (KRUPINSKI et al., 1976; ROBBERS
and FLOSS, 1976) and chanoclavine cyclase (GRÖGER et al., unpublished results) - it was recently demonstrated that the effector-caused increase in alkaloid production rate is accompanied by an increase in the enzyme activity measured in vitro.

Considering these results, it is of interest that under physiologic conditions, the level of free tryptophan in *Claviceps* cultures
rises two-to threefold during the transition from growth phase
to the alkaloid production phase. The increase is much greater
than that of the total free amino acid fraction, indicating the
specific role of tryptophan for the expression of alkaloid metabolism (ROBBERS et al., 1972). Furthermore, experiments by BU'LOCK
and BARR (1968) with tryptophan-treated *Claviceps* cultures revealed
a fluctuation of the internal tryptophan pool during the phase
of alkaloid production (Fig. 6). Alkaloid accumulation under
these experimental conditions shows typical inflections, and
its second differential ($\frac{d^2 alk}{dt^2}$), which is a measure of the rate
of appearance and disappearance of the enzyme(s) that limit the
rate of alkaloid synthesis, resembles the fluctuations of the
internal tryptophan concentration. This result again indicates
that there is a direct relationship between both events. Experiments with protein synthesis inhibitors have shown that the enzyme(s) limiting the rate of alkaloid synthesis are metabolically
labile and thus may undergo rapid fluctuations in concentration.

It must be emphasized, however, that *exogenous* tryptophan and tryptophan analogs stimulate alkaloid metabolism only if they are
added at the time of inoculation or during the first hours of
growth. The restriction of efficiency to a period long before
alkaloid metabolism is expressed makes a direct influence on
this segment of the metabolic process unlikely. Thus, in addition
to the regulatory effects during the alkaloid production phase,
there may be other points of tryptophan control in the early
developmental stages of the cultures, which determine the expressibility of secondary metabolism in the idiophase (cf. the results with *Penicillium cyclopium*, Chap. D1e).

◁

[b] Due to its related chemical structur, CO_2 acts in some plants similarly
to ethylene or interferes with ethylene. Ethylene-enhanced anthocyanin accumulation in fruits of *Vaccinium macrocarpon* and in bracts of *Euphorbia myrsinites* is inhibited by CO_2. Increase of anthocyanin accumulation in *Sorghum vulgare* and corresponding inhibition of anthocyanin accumulation in *Beta vulgaris* is mimicked by CO_2 (CRAKER and WETHERBEE, 1973).

[c] In *Sinapis alba* and *Chenopodium rubrum* cAMP showed no influence on anthocyanin and betalain synthesis, respectively, in spite of the fact that in
S. alba the endogenous level of cAMP is indeed controlled by phytochrome
(JANISTYN and DRUMM, 1975). By AMRHEIN (1974) no cAMP could be found in any
of the higher plants tested.

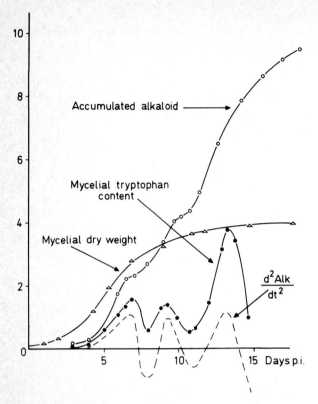

Fig. 6. Alkaloid accumulation and tryptophan content of submerged cultures of *C. purpurea* (after BU'LOCK and BARR, 1968, redrawn). At time of inoculation, nutrient solution was supplemented with 2.5 mM tryptophan. △—△ Mycelial dry weight (10 = 17 g/l), O—O alkaloid accumulated in culture medium (10 = 5 mM), ●—● tryptophan content of mycelium (10 = 5 mM), ——— second differential of accumulated alkaloids ($\frac{d^2 alk}{dt^2}$)

2. Stimulation of Streptomycin Formation in *Actinomyces streptomycini* by the A-Factor

KHOKHLOV et al. (1967) reported the isolation of a highly active substance, called A-factor, that specifically affected the expression of streptomycin production in *Actinomyces streptomycini*. The A-factor was found in the culture broth of certain *Actinomyces* strains (Fig. 7) and has been subsequently purified 2×10^5-fold. One mg of the purest preparations induces the formation of 5×10^4 mg streptomycin in the A-factor-defective mutant 1439. The A factor has the chemical formula $C_{13}H_{22}O_4$, is soluble in organic solvents, heat stable (60°C), and destroyed by acid or alkali treatment. Its chemical structure could be recently defined as 2S-isocapryloyl-3R-oxymethyl-γ-butyrolactone (KLEINER et al., Bioorganitcheskaja Chimia 2, 1442-1447, 1976). It thus shows no structural relation to any part of the streptomycin molecule. In the mutant strain 1439, the transamidinase catalyzing the formation of streptidine, the core of the streptomycin

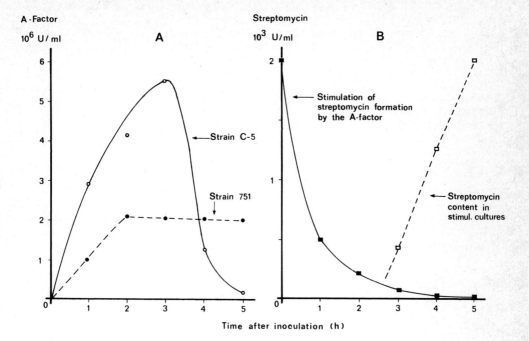

Fig. 7. A-factor content in high- and nonproducing strains of *A. streptomycini* (A) and stimulation of streptomycin formation in nonproducing mutant 1439 (B) (from TOVAROVA et al., 1970, redrawn). A: Amount of A-factor in culture broth determined by adding aliquot of culture broth to cultures of mutant 1439. Strain C-5 is streptomycin high-producing, strain 751 nonproducing. B: Stimulatory effect of A-factor on mutant 1439 tested by adding equal amounts of A-factor at indicated points during culture development. Full curve gives maximal streptomycin content of individual cultures in relation to time of A-factor addition. Dashed line indicates dynamics of streptomycin formation in mutant 1439 when cocultivated with nonproducing, A-factor-containing mutant 751 (calculated from KHOKHLOV et al., 1967)

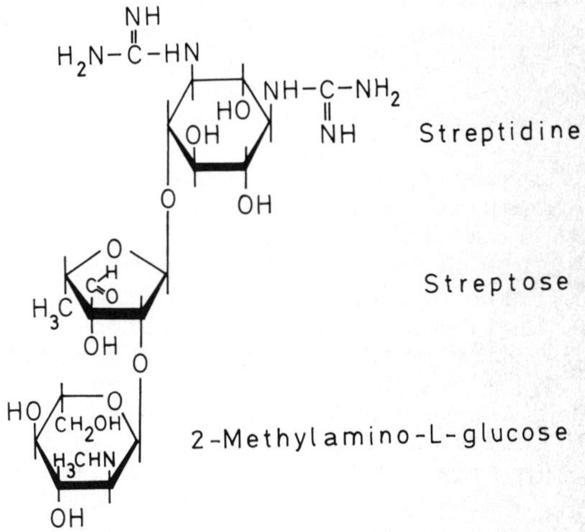

Fig. 8. Streptomycin

molecule (Fig. 8), by transfer of guanidine groups from arginine to streptamin, is measurable only after addition of the A-factor. Thus differential gene expression is evidently involved in its regulatory effect.

The A-factor is excreted equally from a number of *A. streptomycini* strains irrespective of their capability for antibiotic formation (Fig. 7), though the amount in the culture broth of high-producing strains sharply declines at the end of the culture period (strain C-5 in Fig. 7). Its stimulatory activity in the A-factor-lacking mutant 1439 is expressed only if the compound is added at the time of inoculation or shortly thereafter (Fig. 7), i.e., at a period long before streptomycin formation begins. In this respect there is a striking similarity between the action of the A-factor and the stimulation of alkaloid production in *Claviceps* species by tryptophan and in *P. cyclopium* by phenylalanine derivatives (cf. Chap. C1 and D1e). In all three systems the presumed effector must be added in the early trophophase.

Aside from the expression of streptomycin formation, the A-factor also influences expression of other idiophase events. Addition of the effector to A-factor-lacking mutants normalizes sporulation, morphology, and pigmentation of the colonies as well as activities of enzymes of primary metabolism (KHOKHLOV et al., 1973, 1976). Hence the A-factor evidently is not an effector of secondary metabolism itself. Rather, its action resembles that of the hormones of higher plants and animals. It is of interest that in the *Actinomyces* strains at least one further factor with hormone-like properties exists that also influences the expression of the idiophase events (KHOKHLOV et al., 1976; for related experiments with extracts from *P. cyclopium*, cf. Chap. D1e).

3. The Effect of Cinnamic Acids on Anthocyanin Synthesis in *Petunia hybrida*

The flowers of *Petunia hybrida* contain several anthocyanin glucosides that derive from aglyca cyanidin, peonidin, petunidin, malvidin, and delphinidin. During certain developmental stages these compounds are formed in the petals from acetate and malonate, respectively, and from cinnamic acid derivatives, which in their substitution patterns correspond to those of the B-rings of the anthocyanins (HESS, 1967a) (cf. Fig. 2).

Experiments with externally added cinnamic acid derivatives revealed two basic effects on anthocyanin formation (Fig. 9):

1. For some compounds there is a general promotion or inhibition respectively, of overall anthocyanin production. This mechanism was not investigated further.
2. After administration of ferulic acid, 5-hydroxy ferulic acid, and 5-methoxy ferulic acid, there is an increase in formation of those anthocyanins with the corresponding substitution pattern in the B-ring. This specific stimulation can be repressed by inhibitors of gene expression, e.g., by chloramphenicol, actinomycin, and puromycin (HESS, 1967b, 1968).

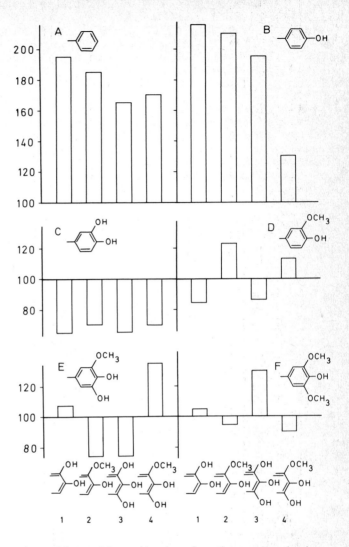

Fig. 9. Influence of cinnamic acids on biosynthesis of anthocyanins in isolated petals of P. *hybrida* (cyanidin type) (from HESS, 1967a). Petals of cyanidin type of P. *hybrida* were fed with acetate-1-^{14}C and indicated cinnamic acids:(A) cinnamic acid,(B) p-coumaric acid,(C) caffeic acid,(D) ferulic acid,(E) 5-hydroxyferulic acid,(F) sinapic acid. Incorporation of radioactivity was used as measure of anthocyanin biosynthesis (incorporation in controls without cinnamic acids = 100). On baseline of figure, substitution pattern of B-ring of types of anthocyanins investigated are shown: (1) cyanidin-3-monoglucoside, (2) peonidin-3-monoglucoside, (3) delphinidin-3-mono glucoside, (4) petunidin-3-monoglucoside

The reversion by chloramphenicol of ferulic acid-stimulated peonidin-3-monoglucoside formation (HESS, 1967b) was taken as evidence for the requirement of differential gene expression. Chloramphenicol is a specific inhibitor of protein biosynthesis in the prokaryotic systems, including those in the organelles of eukaryotic cells (SMILLIE and SCOTT, 1969; PARTHIER, 1970). Thus, the effectiveness of this drug in fact would indicate that these organelles play a decisive role in the induction of anthocyanin formation (cf. Chap. D5).

During petal development, the inducibility of peonidin biosynthesis is greatest at the time when peonidin formation starts. At the same stage, the endogenous ferulic acid concentration increases (Fig. 10). This indicates that the induction by these compounds is a physiologic process and is not due to the high substrate concentrations used in the feeding experiments (HESS, 1968).

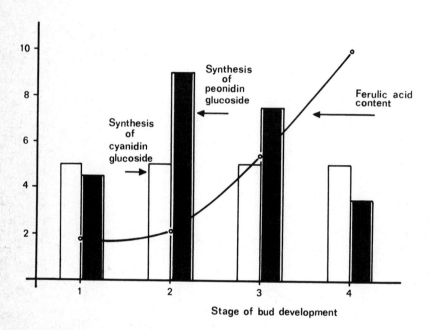

Fig. 10. Rate of peonidin-3-monoglucoside formation and ferulic acid content during development of buds of *P. hybrida* (cyanidin type) (from HESS, 1968, redrawn). Biosynthesis of peonidin-3-monoglucoside (■) in relation to cyanidin-3-monoglucoside (□). For experimental details, cf. Fig. 9. Content of ferulic acid (10 = 1.4 µmol)

4. The Influence of the Light-Phytochrome System on the Regulation of Phenylpropanoid Metabolism in Higher Plants

Phytochrome, a chromoprotein characteristic of higher plants, exists in two spectral forms that are interconvertible by irradiation with far red light (maximum at 730 nm) and near red light (maximum at 660 nm), respectively:

$$\text{Phytochrome P 660} \underset{\text{far red}}{\overset{\text{near red}}{\rightleftarrows}} \text{phytochrome P 730}$$

       (Pr)   *(Pfr)*

physiologically   *physiologically*
inactive   *active*

In the dark Pr is the only form of phytochrome present whereas, with irradiation, an equilibrium concentration of both Pfr and Pr is found, the ratio being dependent on the wavelength used. With white day light, the physiologically active form predominates.

With regard to secondary metabolism the multifold stimulatory effects of phytochrome include:

1. The formation of cinnamic acid derivatives, anthocyanins, and other flavonoids as well as of the enzymes involved in the synthesis of these compounds, e.g., PAL, cinnamic acid 4-hydroxylase, p-coumaric acid:CoA ligase, chalcone flavanone isomerase, UDP-apiose synthetase, apiosyl transferase and glucosyl transferase in many higher plants and corresponding cell cultures (for summaries of the literature, cf. SMITH, 1972, 1973; MOHR et al., 1974; McCLURE, 1975; cf. also DRUMM et al., 1975)
2. The formation of carotenoids, e.g., in ripening tomatoes (THOMAS and JEN, 1975)
3. The synthesis of gibberellic acids by etiolated wheat leaves (COOKE and SAUNDERS, 1975; COOKE et al., 1975)
4. The synthesis of betalains in *Amaranthus* species and *Celosia plumosa* (GIUDICI DE NICOLA et al., 1973)
5. The formation of ethylene (CRAKER et al., 1973).

Early effects of Pfr are observed 15-30 sec after the beginning of irradiation: there is a pronounced increase in the permeability of the protoplasmic membrane (WEISENSEEL and HAUPT, 1974; SCHÄFER, 1974). The significance of this effect for secondary metabolism is supported by experiments with *Brassica oleracea* seedlings. In this plant phytochrome-mediated anthocyanin synthesis is inhibited by Ca^{++} and cholesterol, two compounds that stabilize cell membranes. After treatment with light the amount of free Ca^{++} in the seedling is significantly increased, and administration of EDTA promotes synthesis of anthocyanin in the dark (HATHOUT BASSIM and PECKET, 1975). Furthermore in vitro, the reversion of isolated active Pfr to Pr in the dark is accelerated by Ca^{++} or Mg^{++} ions, an effect that is also inhibited by chelating agents (NEGBI et al., 1975). These results suggest that the primary effect of phytochrome is intimately connected with membrane permeability and the intracellular concentration of divalent ions.

Another early effect detectable by immunocytochemical methods long before any change in gene expression is observed is an alteration of the distribution pattern of phytochrome within the cell. Whereas Pr is generally distributed throughout the cytoplasm, Pfr, within less than 8 min after the onset of saturating red irradiation, becomes associated with discrete regions of the cell. These regions do not appear to be nuclei, plastids, or mitochondria. After phototransformation back to Pr, phytochrome slowly resumes its general distribution. It can be speculated that this light-induced change in the compartmentation of the regulatory active form of the chromoprotein Pfr is due to binding to the physiologically active receptor sites of the cells (MACKENZIE et al., 1975).

The effects of phytochrome on the expression of secondary metabolism can be detected much later. This is best demonstrated for the key enzyme of phenylpropanoid metabolism, PAL, which catalyzes the synthesis of cinnamic acid from phenylalanine (Fig. 1).

The dynamics of the level of PAL and similarly the kinetics of the formation of secondary products derived from cinnamic acid, e.g., that of anthocyanins, show three typical phases after illumination (Fig. 11):

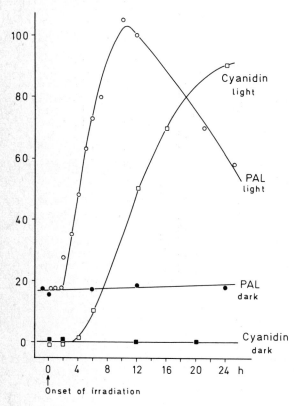

Fig. 11. Kinetics of PAL activity and of anthocyanin accumulation in excised buckwheat hypocotyls during illumination (from AMRHEIN and ZENK, 1970, redrawn). Cyanidin accumulated (100 = 0.4 µmol/g fresh weight), PAL, specific activity (100 = 15 µmol/min/mg)

1. A lag phase, which in buckwheat hypocotyls is about two hours until the increase of PAL activity and about four hours until that of anthocyanin formation

2. A phase of an approximately linear increase of the in vitro measurable enzyme activity and of pigment synthesis, respectively
3. A phase characterized by rapid loss of enzyme activity and cessation of product formation. During this phase, the enzymatic activity measurable in vitro either decreases to the level observed before the onset of irradiation or, after some time, stabilizes at a somewhat higher level.

The duration of the lag phase evidently depends on the plant material used and may even be dissimilar in different cells of the same plant (GILL and VINCE, 1969; BELLINI and MARTINELLI, 1973). However, it is independent of the length of irradiation time and of the quality of light (LANGE et al., 1971).

The existence of a lag phase between the addition of the extracellular stimulator and the appearance of the increased amount of the biologically active protein is a general phenomenon of gene expression. It may be as short as 90 sec in bacteria (KEPES, 1969) but is always much longer in eukaryotic cells. A multiplicity of processes may contribute to the overall effect: (1) uptake and possible intracellular processing of the regulating signal, (2) intracellular transport of the effector and its binding to the sites of gene expression, (3) the time required for transcription, processing of mRNA, and the initiation of translation, (4) proteinogen processing if the product of translation is a proteinogen that after release from the polysomes requires activation etc.

This very superficial list demonstrates that the molecular basis of the lag phases of gene expression is very complex. Also in the case of phytochrome action it is largely unexplained. In experiments with *Sinapis alba* seedlings, it was shown that, after the onset of irradiation with red light, an increasing "capacity" for anthocyanin formation is built up in the lag phase, which is expressed even if the seedlings after the light period are returned to the dark (Fig. 12, curve A). However, much of this capacity is destroyed by 5 minutes of far red irradiation (Fig. 12, curve B). These experiments indicate that the Pfr-evoked capacity for anthocyanin biosynthesis exists in two forms. Both are stable in the dark, but during the experiment an increasing segment becomes resistant to far red light, i.e., Pfr inactivation.

Experiments with inhibitors of gene expression revealed that the stable capacity for anthocyanin formation may be due to the formation of mRNA. Actinomycin D inhibits anthocyanin formation only if added during the lag phase (LANGE and MOHR, 1965; HAHLBROCK and RAGG, 1975). In contrast, synthesis of the enzyme protein, which can be prevented by cycloheximide, starts the moment the lag phase is overcome (HAHLBROCK and RAGG, 1975). This finding agrees with labeling experiments, which also have shown that PAL synthesis is increased after the end of the lag phase (HAHLBROCK and SCHRÖDER, 1975; WELLMANN and SCHOPFER, 1975). Isolation and quantification of PAL-mRNA by in vitro translation and immunoprecipitation of the radioactively labeled enzyme proved

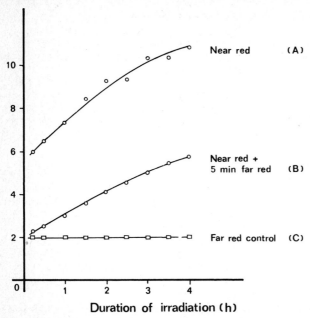

Fig. 12. Test for reversibility during initial lag phase of phytochrome action on seedlings of S. alba (from LANGE et al., 1971, redrawn). S. alba seedlings grown in dark were irradiated with near red light for variable amounts of time (abscissa) and A: returned directly to dark or B: returned to dark after 5 min irradiation with far red light. C: Control was kept in dark except for 5 min irradiation with far red light at indicated times. Anthocyanin content was determined after 24 h

directly that the rate of PAL synthesis in *Petroselinum hortense* cell cultures is limited by the amount of PAL-mRNA, i.e., that the main control point for PAL synthesis is at the transcriptional level (HAHLBROCK, 1977).

The far red labile capacity of anthocyanin formation has not yet been investigated in detail. It appears to be directly connected with the level of active phytochrome and may reflect the early membrane effects and/or redistribution of phytochrome after irradiation, as discussed above.

The second phase after the onset of irradiation, shown in Fig. 11, i.e., the Pfr-mediated increase of PAL activity, is due to a de novo synthesis of the enzymes as shown by density labeling experiments. Though a marked synthesis of PAL was also observed in seedlings of *Sinapis alba* grown in the dark, it was tenfold higher after illumination (ACTON and SCHOPFER, 1975).

The third phase, characterized by a cessation of product formation and a rapid decline of PAL activity, seems to be caused by two independent effects:

1. On the one hand, PAL is strongly inhibited by cinnamic acid derivatives, benzoic acids, and flavonoids (CAMM and TOWERS, 1973; IREDALE and SMITH, 1974; JANGAARD, 1974; ENGELSMA, 1974).
2. On the other hand, density labeling experiments indicate that cinnamic acid represses de novo formation of PAL (JOHNSON et al., 1975; cf. also ENGELSMA, 1968). Because PAL has a high turnover rate (ACTON and SCHOPFER, 1975), lack of de novo formation will cause a rapid decrease of the amount of enzyme present. Recently HAHLBROCK (1976) proposed a general model that explains the dynamics of PAL activity in cell cultures of *P. hortense* by changing rates of enzyme synthesis at constant rates of degradation.

Not in all plants, however, increase of the in vitro measurable activity of PAL after illumination or treatment with other PAL-"inducing" agents is caused by de novo formation. There is evidence that in some organisms it is due to activation of pre-existing, inactive PAL (cf. e.g., ATTRIDGE et al., 1974a, b; CREASY and ZUCKER, 1974). Thus, when seedlings of *Cucumis sativus* grown in the dark are first exposed to low temperature and after some time returned to $25^{\circ}C$, there is an increase of PAL activity that is intensified rather than inhibited by cycloheximide (Fig. 13).

The existence of a thermolabile, high molecular weight compound was recently demonstrated in seedlings of *C. sativus*, which, depending upon the light conditions, reversibly inactivates PAL (ATTRIDGE et al., 1974a, b; FRENCH and SMITH, 1975; JOHNSON et al., 1975). Taking the existence of this inhibitor into consideration, the kinetics of PAL activity shown in Figure 13 may be explained by the following regulatory features (SMITH, 1973; Fig. 14): (1) a continuous synthesis of PAL and PAL-inactivating protein in the dark, (2) the formation of a complex of both proteins that has no PAL activity and that at low temperature is especially stable, (3) the breakdown of a large part of the complex by the temperature shift back to $25^{\circ}C$, and (4) a more rapid turnover of the inactivating protein compared with that of PAL, which leads to higher PAL activity if de novo synthesis of PAL and the inactivator is inhibited by cycloheximide.

A remarkable feature of phytochrome action on plant cells is that its particular effects depend on the developmental stage of the individual cells. This phase dependence is closely connected with the phenomena of competence and determination (cf. Chap. D). The individual cells of a tissue, through prior processes of differential gene expression (formation of regulatory proteins or effectors), are capable of reacting to the same developmental signal with different determined differentiation programs. For instance, after illumination, some epidermal cells in the hypocotyl of mustard seedlings start hair formation whereas the subepidermal cells accumulate anthocyanins.

On the other hand the competence states of the cells change during differentiation. This can be demonstrated by phytochrome-evoked

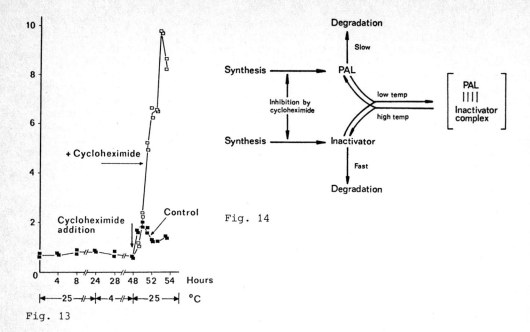

Fig. 13.

Fig. 14

Fig. 13. Effect of transfer to low temperature and cycloheximide treatment on extractable PAL in seedlings of *C. sativus* (from ATTRIDGE and SMITH, 1973, redrawn). 4-day-old gherkin seedlings were kept for 24 h at 25°C, transfered to 4°C for 24 h, and then returned to 25°C. At time of last transfer, one set of plants was treated with cycloheximide (100 μg/ml). Entire experiment was performed in dark. 10 = 5000 dpm/20 hypocotyls · hour

Fig. 14. Mechanisms involved in regulation of PAL level in seedlings of *C. sativus* (from SMITH, 1973, redrawn)

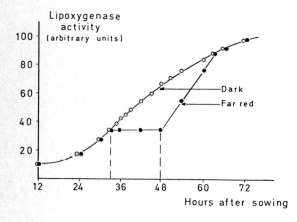

Fig. 15. Increase of lipoxygenase in mustard seedling in continuous dark and under continuous far red light (from OELZE-KAROW and MOHR, 1970). Far red irradiation was started at time of sowing. Repression of lipoxygenase after illumination is detectable between 33 and 48 h only

lipoxygenase repression in *Sinapis alba* seedlings (OELZE-KAROW and MOHR, 1970; SCHOPFER and PLACHY, 1966). Whereas lipoxygenase activity in seedlings grown in the dark increases constantly until 72 h after sowing, illumination between 33 and 48 h stops accumulation of enzyme activity (Fig. 15). Before and after this time interval no light repression occurs. Spectrophotometric measurements of the phytochrome content showed that this temporal pattern of light response is not limited by the amount of active phytochrome. It was also excluded that an increased turnover causes the cessation of enzyme accumulation. The temporal restriction of light action therefore must depend on changes in the Pfr responsive system.

5. The Action of Adrenocorticotrophic Hormone on Corticosteroid Formation in the Adrenal Cortex

The biosynthesis of corticosteroids from cholesterol in the mammalian adrenal cortex involves enzymes partly localized in the mitochondria and partly in the cytoplasm. At first 6 C-atoms of the cholesterol side chain are cleaved off in the mitochondria. The pregnenolone synthesized is released to the cytosol and there transformed to 11-deoxycortisol. The latter compound is 11β-hydroxylated again within the mitochondria to yield cortisol, which is released from the adrenal cells (Fig. 16). Similar reaction sequences lead to the other members of the corticosteroid group (cf. SCHULSTER, 1974).

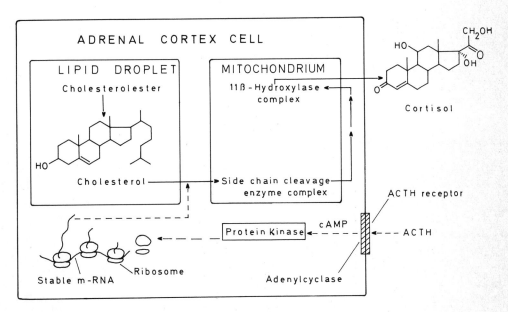

Fig. 16. Hypothetical model of action of ACTH on corticosteroid biosynthesis in adrenal cortex (from SCHULSTER, 1974, redrawn)

ACTH, a linear polypeptide of 39 amino acids, is synthesized by the adenohypophysis. The adrenal gland responds to ACTH with a wide variety of effects, e.g., (1) stimulation of corticosteroid biosynthesis, (2) acceleration of the hydrolysis of cholesterol esters stored in lipid droplets, (3) increase of the flow-rate of blood through the adrenal gland, (4) increase of the adrenal weight, (5) acceleration of phosphate turnover, (6) stimulation of glycogenolysis and glucose oxidation, (7) decrease of the ascorbic acid, lipid, and cholesterol contents.

Some aspects of this complex influence on metabolism in vivo are also observed in in vitro systems. Perfusion of the adrenal gland with a chemically defined medium is frequently used because of its close relation to in vivo conditions. Suspensions of adrenal cells have also been used. Additon of ACTH to the in vitro systems increases the rate of corticosteroid biosynthesis, as in vivo.

ACTH can be coupled to inert polymers (cellulose, agarose, polyacrylamide), preventing its uptake by the receptor cells without loss of biological activity. It was shown that the peptide influences the intracellular concentration of cAMP, which, as a second messenger, mediates the effect of a variety of hormones (JOST and RICKENBERG, 1971; SUTHERLAND, 1972; DRUMMOND et al., 1975). There are two types of experiments that point out that cAMP is involved in increased ketosteroid biosynthesis in the adrenal cortex:

1. Immediately after addition of ACTH the concentration of cAMP is increased (Fig. 17)

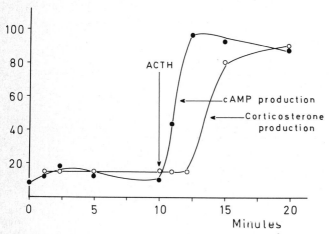

Fig. 17. Time course of cAMP concentration and corticosterone biosynthesis in rat adrenal glands after administration of ACTH (from GRAHAME-SMITH et al., 1967). At time indicated by arrow, ACTH was added to adrenal quarters in vitro (final concentration 1 U/ml). Ordinate: Rate of corticosterone release: 100 = 10 µg/min · 1 g adrenal gland; cAMP: 100 = 15 nmol/1 g adrenal gland

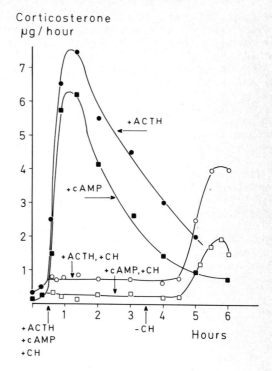

Fig. 18. Kinetics of corticosterone biosynthesis by perfused adrenal glands from hypophysectomized rats after administration of ACTH or cAMP (from SCHULSTER, 1974, redrawn). At indicated time ACTH (64 U/ml) or cAMP (2 mM) was added to perfusion solution. Cycloheximide (CH) (1 mM) was given simultaneously to indicated samples. Cycloheximide administration was stopped after 3 h whereas ACTH and cAMP administration was continued

2. cAMP and its dibutyryl derivative mimic the effect of ACTH (Fig. 18).

ACTH increases cAMP formation by activating adenyl cyclase rather than by inactivating the phosphodiesterase degrading the effector. The peptide thus acts similarly to the other hormones whose effects are mediated by cAMP. It is of interest that aside from ACTH, prostaglandins also increase the formation of corticosteroids in the adrenal cortex. This agrees with the assumption that prostaglandins play an essential role in the signal-chain between extracellular hormone and cAMP formation within the cell (SUTHERLAND, 1972; SILVER and SMITH, 1975).

Experiments with inhibitors of protein biosynthesis (e.g., puromycine and cycloheximide) have shown that the increase of corticosteroid release after administration of ACTH or cAMP probably depends on the de novo synthesis of proteins. The almost immediate and total inhibition by cycloheximide is shown in Fig. 18. The action of cycloheximide is at least partly reversible; however, it takes approximately one hour until corticosteroid release rises again.

Using in vitro systems it has been shown that the latency period between administration of the effectors and the beginning of the increase of corticosteroid biosynthesis (cf. Fig. 17) is comparable to the average transit time (1-2 min) required for translation in eukaryotic cells (FAN and PENMAN, 1970; VAUGHAN et al., 1971). It is therefore postulated that the lag period represents the time required for the translation of protein(s) required for increased corticosteroid formation. If cycloheximide is added to cells undergoing fully induced corticosteroid biosynthesis, release of the corticosteroids falls linearly with a $T_{1/2}$ of about 10 min, a process that may indicate the rapid turnover of the rate-limiting protein.

The mechanism whereby the induced protein(s) acts on corticosteroid formation is at present unknown. Circumstantial evidence indicates that an increased accessibility of cholesterol in the mitochondria may cause the higher production rates.

In recent years a steroid carrier protein (SCP) was isolated from many mammalian cells including those of the adrenal gland (KAU and UNGAR, 1973; MAHAFFEE et al., 1974; DEMPSEY, 1974). It was shown to have specific affinity for cholesterol and to stimulate cholesterol side chain cleavage 5- to 10-fold (KAU and UNGAR, 1973). In fact, no significant change of the activity of enzymes involved in the transformation of cholesterol into corticosteroids was observed after ACTH administration (KARABOYAS and KORITZ, 1965; BOYD et al., 1971; MAHAFFEE et al., 1974). Instead there was a marked depletion of cholesterol from the lipid droplets, which can be prevented by simultaneous administration of cycloheximide (DAVIS and GARREN, 1968; KAU and UNGAR, 1973; MAHAFFEE et al., 1974). It can be speculated that a metabolically unstable SCP is required both for the transport of cholesterol to the mitochondria and for its conversion to pregnenolone and that the regulatory influence of ACTH on steroidogenesis is brought about by the increased formation of this protein, which then increases the mitochondrial steroid precursor pool (MAHAFFEE et al., 1974; DEMPSEY, 1974). In contrast to the inhibitors of translation, those of RNA biosynthesis have little or no effect on ACTH-induced steroidogenic response. Actinomycin D, for instance, in doses that inhibit ^{14}C-adenine and ^{14}C-uridine incorporation into RNA by about 95%, decreases ketosteroid release only by about 20% (SCHULSTER, 1974). This result indicates that ACTH stimulation does not directly require de novo synthesis of RNA and that formation of the presumed rate limiting proteins (SCP) must depend on the existence of preformed RNA.

In summary, corticosteroid formation induced by ACTH in adrenal cortex cells may be regulated as follows (SCHULSTER, 1974, Fig. 16): The peptide hormone binds to the outer side of the protoplasmic membrane and activates adenylcyclase, an enzyme located at the inner side. The cAMP formed leads to specific phosphorylation of ribosomal proteins, either by activating protein kinases associated with ribosomes or by cAMP-directed translocation of protein kinases to the ribosomes (JUNGMANN et al., 1975). The phosphorylation of ribosomal proteins causes the translation of stored, stable mRNA(s). SCP is synthesized, which by an as yet unknown mechanism facilitates the translocation of cholesterol from the storage site to the mitochondria. In this way the precursor supply of the enzymes of corticosteroid biosynthesis is inreased, and the initial steps of cholesterol conversion are stimulated.

D. Phase Dependence of Secondary Metabolism and the Organization of Differentiation Programs

Phase dependence of secondary product formation in microbial cultures is a well-known phenomenon in industrial microbiology (DEMAIN, 1968, 1972). According to a suggestion by BU'LOCK (BU' LOCK et al., 1965), the main developmental phases of such cultures are designated as *trophophase* (phase of rapid, exponential increase of cell numbers) and *idiophase*, respectively. The latter is frequently characterized by a slowing down of growth or by its specialization (cf. Chap. D1b) and particularly by production of special proteins including the enzymes forming secondary metabolites. Thus the expression of secondary metabolism is part of a series of profound changes that occur during the transition from trophophase to idiophase and that proceed as a differentiation program leading to multifold specialization of the individual cells.

Though these basic phenomena are also observed in higher plants and animals, the highly specialized nature of the cells of these organisms at very early stages of development precludes a simple application of the pertinent terms tropho- and idiophase. Although it is almost trivial to state that secondary products are formed by some specialized cells in all developmental stages of higher plants and animals, in only a very few cases the formation of the cells involved and their subsequent specialization was investigated to characterize the overall process as a sequence of trophophase-idiophase-like periods. Phase dependent formation of specialization proteins in developing animal erythroblasts and myoblasts, respectively (HOLTZER and ABBOTT, 1968; HOLTZER et al., 1973; WEINTRAUB, 1975) and in developing leguminous seeds, represent interesting model systems. These seeds develop in three phases. A phase of active cell division (trophophase) is followed by a determinative phase defined by endomitoses, proliferation of the endoplasmic reticulum, and an increase of the number of membrane-bound ribosomes, and then by the specialization phase, during which the synthesis of storage proteins occurs (BRIARTY et al., 1969; BOULTER et al., 1972; MÜNTZ et al., 1972).

Before characterizing further the expression of secondary metabolism as an integral part of differentiation programs, we will briefly outline the regulatory principles governing such programs as they presently can be deduced from experimental investigations, especially -

1. On the regulation of enzymatic adaptation in bacteria (RICHMOND, 1968; HUANG, 1972; GOLDBERGER, 1974; HUA and MARKOVITZ, 1974; ENGELSBERG and WILCOX, 1974; MAGASANIK et al., 1974; RICKENBERG, 1974; CALHOUN and HATFIELD, 1975; FOOR et al., 1975; STEPHENS et al., 1975; MAGASANIK, 1977)
2. On the programmed expression of the viral genetic material in the corresponding host cells (SZYBALSKI et al., 1970; BAUTZ, 1972; HUANG, 1972; LOSICK, 1972; OTA et al., 1972; SUGIYAMA et al., 1972; DUFFY et al., 1975; MEYER et al., 1975; MAILHAMMER et al., 1975; PERO et al., 1975; SZYBALSKI, 1977)

3. On the interaction of steroid hormones with their target cells (JENSEN et al., 1969; TOMKINS et al., 1969; JENSEN and DE SOMBRE, 1973; O'MALLEY and MEANS, 1974; COHEN and HAMILTON, 1975; WOO and O'MALLEY, 1975) and
4. On the possible function of chromosomal RNA and nonhistone proteins in the eukaryotic chromatin (HUANG and SMITH, 1972; ELGIN and WEINTRAUB, 1975; PAUL and GILMOUR, 1975; TSAI and O'MALLEY, 1977).

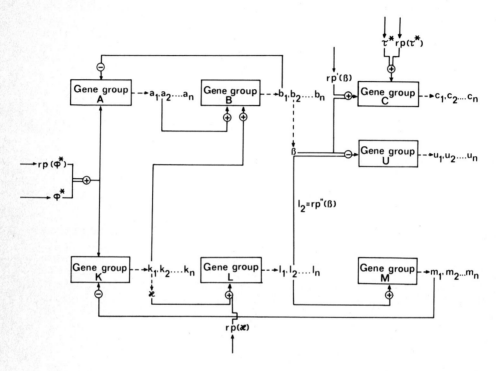

Fig. 19. Hypothetical scheme of coordination of individual steps of differential gene expression into programs. A, B, C ....: in squares, expression of gene groups A, B, C; ---→ a_k, b_k, c_k ...: formation of regulatory RNAs and of proteins (enzymes, structural and regulatory proteins) coded by gene groups A, B, C ....; ---→ α, β, γ ....: formation of regulatory effectors by enzymes a_k, b_k, c_k ....; $rp_{(\alpha)}$, $rp_{(\beta)}$ ....: receptor proteins required for the regulatory interaction of effectors (α, β ....) with corresponding gene expression chains; φ*, τ*....: effectors coming from outside the cell; —⊕—→ / —⊖—→ positive and negative influence, respectively, of regulatory proteins and effectors of gene expression

The main features of the scheme shown in Fig. 19 are:

1. Processes of differential gene expression are interrelated through the action of regulatory RNAs, proteins, or effectors, which themselves are formed directly or indirectly by prior processes of gene expression.
2. The effectors involved are either formed within the cell (β, κ) or come from outside (φ*, τ*) and then serve as agents for

the intercellular coordination of differentiation processes. They act by modifying the activity of regulatory proteins.
3. There may be multiple regulation of individual gene groups (cf. gene groups B and C). In fact, this is the rule rather than the exception. Regulation is positive or negative and may concern any one of the partial processes of differential gene expression (cf. Chap. A). The same regulatory RNAs, proteins, or effectors may exert positive or negative actions on one or on several gene groups (cf. l_2).
4. Subprograms may proceed independently of each other. Their further progress, however, usually seems to require regulatory interaction with other subprograms of the same cell or with extracellular effectors (e.g., aside from effectors β and τ^* receptor protein $rp'(\beta)$ as well as receptor protein $rp(\tau^*)$ is necessary for the expression of gene group C).
5. The competence of the cell to respond to the action of extracellular effectors, e.g., ϕ^* and τ^*, is dependent on the presence of the corresponding receptor proteins - $rp(\phi^*)$ and $rp(\tau^*)$.
6. The effectors, e.g., ϕ^*, may trigger a sequence of differential gene expression determined by the regulatory features of the overall differentiation program.

The scheme in its totality is highly hypothetical. With the exception of a few simple examples, the exact functioning of such programs is practically unknown. This is especially true of the intercellular linkages between differentiation processes of individual cells, leading to new characteristics through the functional cooperation of cells within the developmental programs of tissues, organs, and organisms. The only results thus far are limited to a description of the sequence of events within the differentiation programs and to a preliminary analysis of the regulatory interactions by watching the influence of environmental factors or of mutations on the procedure of the programs. These limitations are especially true with respect to the expression of secondary metabolism and to its regulatory integration into such programs. There is, however, no doubt that integration exists, not only in the systems treated on the following pages but also in the systems discussed for special reasons in Chaps. B and C.

1. Integration of Alkaloid Metabolism into the Developmental Program of *Penicillium cyclopium*

a) Growth, Conidia, and Alkaloid Formation

The development of emerged cultures of *P. cyclopium* proceeds in three phases (NOVER and LUCKNER, 1974; LUCKNER, 1977; Fig. 20). After inoculation the conidia spread over the surface of the nutrient solution need about 12 h to overcome their dormancy and to form a germ tube (germination phase). Subsequently (12-72 h p.i.) there is a rapid increase of mycelial dry weight and, concomitantly, of the protein content (trophophase). The incorporation rates of radioactive amino acids and of $^{32}P_i$ into high molecular weight material (proteins and nucleic acids) show maximal values. The end of the trophophase is characterized by a pronounced decrease of the rate of formation of mycelial dry mass, nucleic acid, and protein.

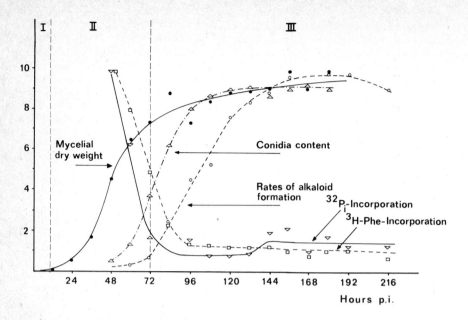

Fig. 20. Dynamics of growth and specialization processes during development of emerged cultures of *P. cyclopium* (from NOVER and LUCKNER, 1974, redrawn). Cultures were grown on nutrient solution (NL I) containing 5% glucose, 0.12% NH_4^+, and 0.025% phosphate. Beginning 48 h p.i. culture broth was replaced every 12 h with nutrient solution (NL II) containing only 20% of original carbon and nitrogen amounts and 2% of phosphate content, respectively. I: Germination phase, II: Trophophase, III: Idiophase. Ordinate: Mycelial dry weight: 10 = 5.4 mg/cm^2 culture area; conidia content: 10 = 6.5 mg conidia/cm^2 culture area; rate of cyclopenin-cyclopenol excretion by the hyphae: 10 = 12 µg/hour · cm^2 culture area; ^{3}H-phe incorporation into TCA insoluble material (rate of protein synthesis): 10 = 100% corresponding to maximum incorporation of ^{3}H-phe 48 h p.i.; $^{32}P_i$ incorporation into nucleic acid fraction (rate of nucleic acid synthesis): 10 = 100% corresponding to maximum incorporation of $^{32}P_i$ 48 h p.i.

It should be noted that the decreased protein synthesis, i.e., the diminuation of the incorporation of radioactive amino acids into the protein fraction, is equally pronounced for leucine and alanine as well as for the amino acids phenylalanine and methionine, which are precursors of the alkaloids formed at later stages of the mould's development. This result indicates that the expression of alkaloid metabolism is not connected to a selective change in the uptake or conversion of the precursors, as was found by KAPLAN et al. (1969) in saprophytic cultures of *Claviceps*. In *Claviceps* the incorporation of the alkaloid precursor tryptophan into protein was decreased in the phase of alkaloid production whereas that of leucine remaind unchanged.

In emerged cultures of *P. cyclopium* beginning about 48 h p.i., the growth processes become more and more concentrated on the surface of the mycelium where, after formation of the penicilli (48-56 h p.i.), the detachment of conidiospores begins. This specializa-

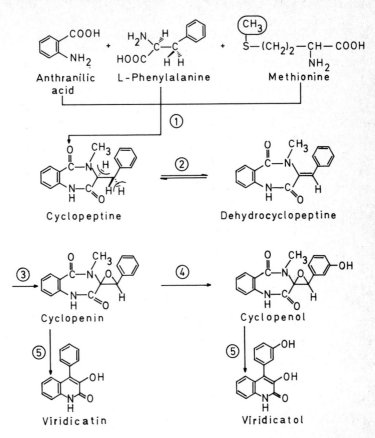

Fig. 21. Formation of the alkaloids of the cyclopenin-viridicatin-group by
P. cyclopium. (1) Cyclopeptine synthetase complex (hypothetic; presumably
catalyzing activation of anthranilic acid and phenylalanine, formation of
peptide bonds, and methylation; intermediates evidently are covalently
bound to enzyme complex; FRAMM et al., 1973); (2) cyclopeptine dehydrogenase
(ABOUTABL and LUCKNER, 1975; ABOUTABL et al., 1976; (3) dehydrocyclopeptine
epoxidase (NOVER and LUCKNER, 1969; VOIGT and LUCKNER, in press); (4) cyclopenin m-hydroxylase (NOVER and LUCKNER, 1969; RICHTER and LUCKNER, 1976); (5)
cyclopenase (WILSON et al., 1974; WILSON and LUCKNER, 1975)

tion growth within 48 h leads to a quantity of conidia that
equals or even surpasses the mycelial dry weight. It is followed,
after a 12-hour delay, by the synthesis of benzodiazepine alkaloids (increasing between 60-168 h p.i.). These alkaloids (cyclopeptine, dehydrocyclopeptine, cyclopenin, cyclopenol) are derivatives of the cyclic peptide of L-phenylalanine and anthranilic
acid (cf. Fig. 21). They are released by the hyphae into the
culture medium. Conidia formation and alkaloid production are
part of a number of specialization processes characteristic of
the third phase of the mould's development, the idiophase. Because
of the almost complete lack of cell divisions in the idiophase
hyphae, the culture area can be used as an equivalent of the
cell number. It is a much more reliable reference base than the
dry weight or the protein content, both of which are severely in-

fluenced by the nutrient conditions. In the following discussion of the regulatory features of alkaloid formation and other idiophase characteristics, therefore, most of the experimental data are calculated per cm^2 culture area.

The beginning of alkaloid formation is characterized by the simultaneous, increasing release of the intermediates and end products of benzodiazepine alkaloid formation, i.e., of cyclopeptine, dehydrocyclopeptine, cyclopenin, and cyclopenol, into the culture medium (Fig. 22). This result indicates coordinated formation of

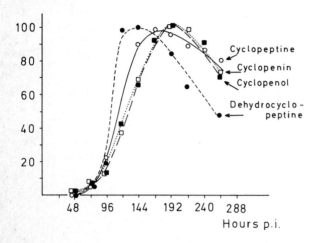

Fig. 22. Dynamics of benzodiazepine alkaloid excretion by cultures of *P. cyclopium* (from FRAMM et al., 1973, redrawn). Cultures were grown as described in Fig. 20. Ordinate: Cyclopeptine: 100 = 21 ng/hour · cm^2 culture area; dehydrocyclopeptine: 100 = 4.4 ng/hour · cm^2 culture area; cyclopenin: 100 = 10.4 µg/hour · cm^2 culture area; cyclopenol: 100 = 4.4 µg/hour · cm^2 culture area

the corresponding enzymes (Fig. 21, Nos. 1-4). However, the activities of the in vitro measurable enzymes cyclopeptine dehydrogenase (CD) and dehydrocyclopeptine epoxidase (DE) have no direct correlation to the dynamics of the rates of alkaloid excretion. That discrepancy is least pronounced in batch cultures (cf. the control curves of Fig. 26A and B). It is most prominent in cultures where mycelia after submerged precultivation are transferred to emerged conditions and continuously supplied with nutrient solution (Fig. 23B).

Because of the immediate inhibitory effect of 5-fluorouracil and cycloheximide on the increased rates of alkaloid formation in cultures in which nutrient solutions are replaced (cf. Chap. D1c), there must be a protein that limits the rate of cyclopenin-cyclopenol formation, which is built up considerably later than CD and DE. Considering the simultaneous increase of all benzodiazepine alkaloids shown in Fig. 22, it is unlikely that the cyclopeptine synthetase complex (Fig. 21, No. 1), which is not yet measurable in vitro, is a good candidate in this respect.

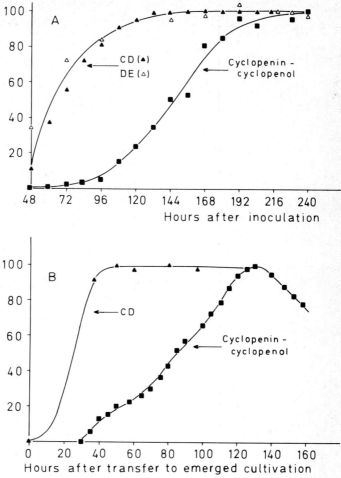

Fig. 23. Alkaloid production, cyclopeptine dehydrogenase, and dehydrocyclopeptine epoxidase activities by different methods of cultivation (EL KOUSY et al., unpublished results). A: Discontinuous exchange of the nutrient solution (cf. Fig. 20); B: Continuous exchange of nutrient solution: Mould was precultivated for 48 h under submerged conditions on rotatory shakers. Globular colonies formed were spread to monolayer on glass plates covered with filter paper and dialysis membrane. Culture was covered with glass plate to prevent extreme evaporation, kept in a laminar flow box and supplied continuously via filter paper with nutrient solution, using NL I during first 12 h of emerged cultivation and later on diluted NL I containing only 15% of original nutrient concentration (NOVER and MÜLLER, 1975). Ordinate: Cyclopeptine dehydrogenase activity of hyphae (▲): 100 = 2.4 (A) and 8.0 (B) mU/cm^2 culture area; dehydrocyclopeptine epoxidase activity of hyphae (△): 100 = 5.1 (A) µU/cm^2 culture area; rate of cyclopenin-cyclopenol excretion by hyphae: 100 = 10.0 (A) and 23.0 (B) µg/hour · cm^2 culture area

One may speculate that the intracellular channeling of precursors to the site of alkaloid synthesis requires a specific protein limiting the rate of production. In agreement with this idea, separated channeling of phenylalanine to the sites of protein and alkaloid synthesis, respectively, has been demonstrated in *P. cyclopium* (NOVER and MÜLLER, unpublished).

b) Formation of the Enzymes of Alkaloid Metabolism during Conidiation

The trophophase-idiophase concept discussed previously does not imply that cell division and cell specialization, e.g., expression of secondary metabolism, are mutually exclusive. There are well-known examples for the direct interdependence of both processes, cf. quantal mitosis during animal erythrogenesis and myogenesis (DIENSTMAN and HOLTZER, 1975; WEINTRAUB, 1975), and the formation of milk proteins in the mammary gland (LOCKWOOD et al., 1967). Conversely, in *Hydra* (BURNETT, 1968), insects (LAWRENCE, 1975), and plants (FOARD, 1970; TRANTHANH VAN et al., 1974; MEINS, 1975), cell specialization evidently also proceeds without cell division.

As far as expression of secondary metabolism is concerned, many examples in this book demonstrate the separation of cell division and chemical specialization. In some systems, however, both processes may occur simultaneously, as shown in plant cell cultures (cf. CONSTABEL et al., 1971) and cultures of microorganisms (cf. DEMAIN, 1972). But the direct regulatory interrelation between cell division and specialization was not proved in any of these examples by experiments with inhibitors of DNA synthesis or function, as was done in the animal systems mentioned previously. In contrast, experiments with *P. cyclopium* demonstrate that the enzymes of benzodiazepine biosynthesis are formed as constitutive parts of nascent conidia and that their activities do not change during maturation.

Conidia detachment in *P. cyclopium* proceeds by repeated unequal division (quantal mitosis) of the apical cells of the penicilli, the so-called phialides, a process taking about two hours per conidium (FALK and WOLLMANN, 1974). After their detachment, the conidia undergo a ripening process that lasts 12-24 h.

In terms of the time of their expression during spore formation and ripening, four groups of events can be distinguished (NOVER and LUCKNER, 1974; cf. Fig. 24): (1) Increase of phenol oxidase activity, which precedes the increase of conidia number by about 2-4 h; (2) formation of conidia, which coincides with the dynamics of the activities of CD and DE involved in the formation of cyclopenin and cyclopenol (EL KOUSY, unpublished) and with the increase of the autoregulated invertase activity; (3) formation of the green pigment, which is characteristic of the ripe conidia and which occurs 4-6 h after spore detachment; (4) appearance of cyclopenase activity, which occurs 12-24 h after conidia detachment, and accumulation of the benzodiazepine alkaloids, cyclopenin and cyclopenol, which lasts until 200 h after detachment.

The formation of phenol oxidase is evidently part of the development of the penicilli. The specific activity of the enzyme is very high in the mixture of aeral hyphae, penicilli, and young conidia obtained by brushing off the upper layer of emerged grown mycelia 48-54 h p.i. In the following hours with the in-

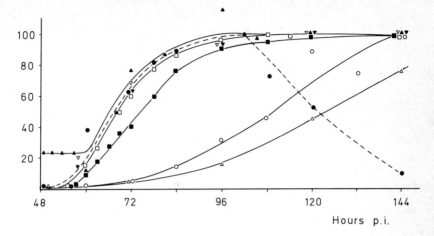

Fig. 24. Biochemical events connected with detachment and ripening of conidiospores of *P. cyclopium* (from NOVER and LUCKNER, 1974, redrawn). Cultures were grown as described in Fig. 20. Values are calculated for 1 cm^2 culture area. ▲: Phenol oxidase activity (arbitrary units); □: number of conidia (100 = 1.7 · 10^8); ▼: CD activity (100 = 5.0 mU); ∇: DE activity (100 = 2.5 µU); ●: invertase activity (100 = 54 mU); ■: pigment amount (arbitrary units); ○: cyclopenase activity (100 = 100 mU); Δ: cyclopenin-cyclopenol accumulation (100 = 27 µg, reached 240 hours p.i.)

creasing amount of conidia, the specific activity of phenol oxidase in this fraction rapidly drops to a constant value. In the meantime, a similar level of phenol oxidase activity is reached also by the hyphae, i.e., later on the enzyme is present in all parts of the late idiophase cultures. It is likely that, in the conidia, phenol oxidase is involved in the biosynthesis of the green pigment of the cell wall (JICINSKA, 1968; CLUTTERBUCK, 1969; MARTINELLI, 1972). The melanine-like structure and the presence of phenolic groups in the pigment (unpublished results), the exoenzyme nature of the phenol oxidase, the high specific enzyme activity in the spore-forming cell fraction, and the preceding of enzyme accumulation in the spores compared with that of the pigment (Fig. 24) are in agreement with this assumption.

The almost coincident dynamics of the activities of CD and DE (cf. Fig. 21, Nos. 2 and 3) with those of conidia detachment, i.e., the constancy of enzyme activity during spore maturation, demonstrates that both enzymes are constitutive proteins of the conidiospores. Because accumulation of cyclopenin and cyclopenol starts immediately after detachment of the spores, the other enzymes involved in the biosynthesis of these alkaloids must be constitutive, too. The same holds true for invertase. Whereas this enzyme in the hyphae of *P. cyclopium* is sucrose-inducible, its apparently transient formation during conidiation is independent of the sucrose concentration of the nutrient solution (ROOS, unpublished results). The formation of the enzymes for cyclopenin-cyclopenol biosynthesis as well as that for invertase thus seems to be part of the process of differential gene expression during the cell division cycle in the phialides.

Though inhibitors of DNA synthesis (ethidium bromide at 200 µg/ml and hydroxyurea at 500 µg/ml) have pronounced effects on the rate of conidia formation, they do not influence the specific activities of the mentioned "constitutive" proteins if calculated on the basis of conidia dry weight. The same was found for the increasing alkaloid formation rate in hyphae. Evidently no DNA synthesis, e.g., in a process of gene amplification (BELL, 1971), is required during cell specialization in *P. cyclopium*.

Clearly separated from the bulk of the "early" differentiation characteristics in the conidia is the formation of the last enzyme of alkaloid metabolism (cf. Fig. 21, No. 5), cyclopenase. Its complete regulatory independence from the remaining subprocesses of alkaloid metabolism is indicated by its restriction to the conidiospores, whereas cyclopenin-cyclopenol biosynthesis proceeds in both hyphae and conidia. In addition, the enzyme is compartmentalized in the conidia themselves. Despite the high activity of cyclopenase (17 U/g dry weight; WILSON et al., 1974), there is very little or no transformation of cyclopenin and cyclopenol to the corresponding quinoline derivatives. The enzyme and its substrates thus must be separated from each other. The enzyme is part of the inner side of the cell wall-bound membrane fraction (WILSON and LUCKNER, 1975) whereas the alkaloids are stored outside the cell membrane in the cell wall (NOVER and LUCKNER, 1974).

As can be expected, the formation of conidia under the influence of inhibitors of gene expression such as fluorouracil and cycloheximide is sharply reduced (EL KOUSY et al., 1975). The amount of the constitutive enzyme invertase is reduced correspondingly, i.e., its specific activity is essentially unchanged. However, the increasing specific activity of cyclopenase during conidia maturation cannot be inhibited. The same effect was detected when isolated immature conidia with low cyclopenase activity were incubated in vitro under submerged conditions with inhibitors of protein synthesis totally inhibiting the incorporation of ^{3}H-phenylalanine into proteins (Fig. 25). From this result the tentative conclusion can be drawn that the increased activity of cyclopenase is due to the conversion of a biologically inactive proteinogen, presumably formed together with the other enzymes of alkaloid metabolism as a constitutive component of the quantal cell cycle during conidia detachment. Transformation of this protein into cyclopenase would be one of the few examples of proteinogen processing thus far known in secondary metabolism (cf. the activation of phenol oxidase in insects, Chap. D8).

c) Influence of Inhibitors of Gene Expression on Hyphal Cyclopenin-Cyclopenol Formation

In order to select compounds to be used as inhibitors of gene expression in experiments with *P. cyclopium*, a number of antibiotics (actinomycin D, puromycin, and cycloheximide) and analogs (8-azaguanine, 6-azauracil, 8-azaadenine, 5-fluorouracil, p-fluorophenylalanine, and β-thienylalanine) were tested for their effects

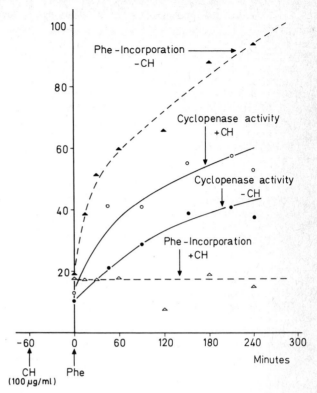

Fig. 25. Cycloheximide-independent increase of cyclopenase activity in conidiospores of P. cyclopium (from EL KOUSY et al., 1975, redrawn). 72 h p.i., conidiospores of batch cultures were brushed off and resuspended in water (2 g fresh cells/ 50 ml). To suspension, 100 µg/ml CH were added, 60 min later experiment was started by adding 0.05 µmol/ml of D,L-phenylalanine-[G-^{3}H] (= 0.1 µC). Cyclopenase activity and phenylalanine incorporation were determined according to NOVER and LUCKNER (1974). Ordinate: Cyclopenase activity: 100 = 1.6 mU/mg fresh conidiospores; phenylalanine incorporation: 100 = 2 · 10^4 cpm/mg protein

on growth and on different features of idiophase specialization. Only cycloheximide and 5-fluorouracil were found suitable, most of others even at concentrations of 100 µg/ml being only slightly active or inactive. The application of p-fluorophenylalanine, though a good inhibitor of gene expression in our test organism, was evidently complicated by its direct influence on alkaloid metabolism (cf. Chap. D1e).

5-Fluorouracil was shown to be incorporated into RNAs and thus to inhibit its processing or biologic function in eukaryotic cells (INGLE, 1968; WALTON et al., 1970; WILKINSON and PITOT, 1973) and bacteria (HOROWITZ et al., 1960). It appears to have similar effects in P. cyclopium. At concentrations of 50-100 µg/ml it exerts no direct effect on protein synthesis (^{3}H-phe-incorporation). Soon after its addition, however, the incorporation of ^{32}P$_i$ into mature rRNA is selectively stopped. This inhibition is only slowly reversible even after adding an excess of uracil.

Cycloheximide is described as an inhibitor of initiation, elongation, and chain termination in eukaryotic protein biosynthesis (PESTKA, 1971). Because of its excellent penetration characteristics, it is frequently used for in vivo experiments. However, many investigations have shown that cycloheximide interferes with processes other than protein synthesis (for a summary of the literature, see McMAHON, 1975). Use of this drug therefore requires an especially careful examination of possible side effects.

In *P. cyclopium*, cycloheximide at concentrations of 10-100 µg/ml immediately inhibits protein synthesis to 90%. But only with 100 µg/ml the effect is stable for 24 h, whereas 95% of the original incorporation rate of ^{3}H-phenylalanine into proteins is reached 24 h after the addition of 10 µg/ml. Evidently cycloheximide is inactivated slowly by the fungus. At high concentrations of the drug, the total capacity for protein synthesis returns to normal about 10 h after its removal.

The effects of cycloheximide and fluorouracil on the expression of alkaloid metabolism in batch cultures are shown in Fig. 26. Shortly after addition there is an almost complete inhibition of the increase of alkaloid formation rate in the hyphae (Fig. 26A I, A II). One of the enzymes of alkaloid biosynthesis, cyclopeptine dehydrogenase (Fig. 21, No. 3), shows almost identical dynamics (Fig. 26B I, B II). Furthermore both types of curves closely resemble those showing the action of the inhibitors on β-galactosidase induction (cf. Fig. 26C I and C II), despite of the significant differences in the time kinetics of enzyme induction. The following conclusions can be drawn from these results (EL KOUSY et al., 1975):

1. The increased rates of alkaloid formation in the hyphae of idiophase cultures of *P. cyclopium* are dependent on permanent RNA and protein synthesis, i.e., are due to a process of chemical differentiation of the hyphae.
2. The striking similarities between the dynamics of the two "induced" enzymes belonging to totally different metabolic chains demonstrate that nonspecific effects of the inhibitors, besides their action on gene expression, are negligible. There is for instance no direct influence of either cycloheximide or fluorouracil on the synthesis of the alkaloids themselves.

As discussed previously (cf. Chap. D1a), an as yet unknown principle limits the rate of alkaloid biosynthesis in vivo. Experiments with cycloheximide and 5-fluorouracil indicate that the synthesis of a protein is involved. Even in the period in which the activities of cyclopeptine dehydrogenase and dehydrocyclopeptine epoxidase reached their peak further increase of the in vivo alkaloid production rates is immediately stopped after the addition of both drugs (LUCKNER, 1977).

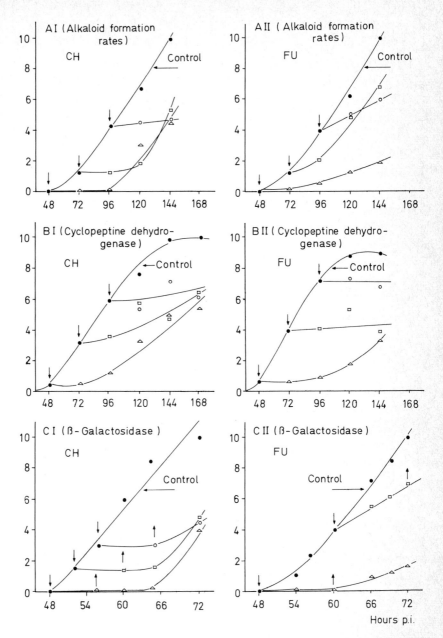

Fig. 26. Influence of cycloheximide and 5-fluorouracil on rates of alkaloid formation and increase of CD and induced β-galactosidase activities (from EL KOUSY et al., 1975). Cultures were grown under batch conditions on NL I. A and B: Inhibitors (100 µg/ml) were added to cultures at time indicated by arrows. C: For induction of β-galactosidase, culture disks 48 h p.i. were transferred to diluted nutrient solution containing arabinose instead of glucose. 8 h (CH) and 12 h (FU), respectively, after addition of inhibitors, culture broth was replaced by inhibitor-free, diluted arabinose nutrient solution (ININGER and NOVER, 1975). Ordinate: A: Rates of alkaloid formation (= excretion of cyclopenin and cyclopenol by hyphae into culture medium), 10 = 4 µg/h · cm^2 culture area; B: CD activity in hyphae: 10 = 5 mU/cm^2 culture area; C: β-galactosidase activity in hyphae: 10 = 120 mU/cm^2 culture area

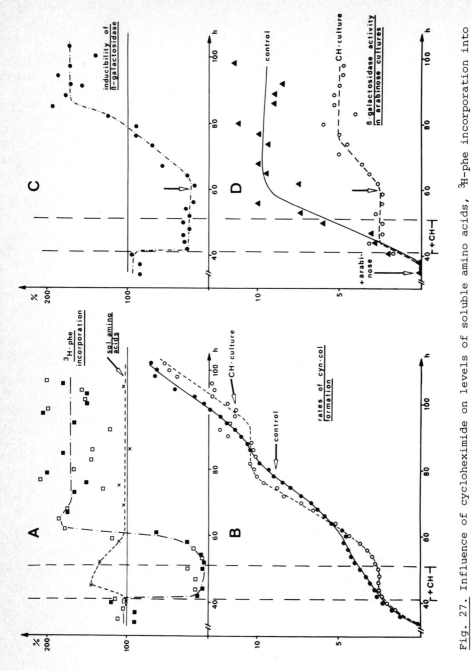

Fig. 27. Influence of cycloheximide on levels of soluble amino acids, ^{3}H-phe incorporation into proteins, rates of cyclopenin-cyclopenol formation, and induction of β-galactosidase in cultures of *P. cyclopium* with continuous nutrient supply (from NOVER and MÜLLER, 1975). Submerged cultures

Detailed analysis of the action of cycloheximide on the rate of alkaloid biosynthesis in cultures with continuous nutrient supply showed a three-phasic response (Fig. 27B) that does not seem to parallel corresponding changes of the in vitro activity of, e.g., cyclopeptine dehydrogenase: (1) After addition of the drug, the alkaloid formation rate stops immediately at the level reached before (primary inhibition). (2) Removal of cycloheximide results in an almost instantaneous 20-30 h increase in alkaloid formation rate surpassing that in the control cultures (superinduction). (3) After a second stoppage of the alkaloid formation rate (secondary inhibition), an increase essentially identical with that in the control cultures follows.

This type of 40-50-hour oscillation of alkaloid formation rates compared to those in the control cultures is found irrespective of the time of cycloheximide addition and hence of the rate of alkaloid formation reached before (NOVER and LUCKNER, 1976). However, the "superinduction" effect is most pronounced if cycloheximide is added 40-60 h after transfer.

Whereas "primary inhibition" correlates with the almost complete inhibition of ^{3}H-phenylalanine incorporation into trichloroacetic acid-precipitable material, secondary inhibition occurs despite an even increased protein biosynthesis. "Superinduction" also deserves special notice. Because cycloheximide interferes neither with alkaloid production per se nor with the excretion of cyclopenin-cyclopenol into the culture solution, the following possible explanations of the latter two processes may be considered:

1. Inhibition of protein synthesis, which leads to a transient increase of the levels of free amino acids in the hyphae (Fig. 27A) including those of the precursor amino acids phenylalanine and methionine, may stimulate alkaloid formation. A direct substrate effect, however, is unlikely because of the rapid normalization of this effect. But it cannot be excluded that the brief increase in the level of phenylalanine is related to the subsequent superinduction phenomenon because phenylalanine evidently plays an as yet ill-defined regulatory role in the expression of alkaloid metabolism in *P. cyclopium* (cf. Chap. D1e).

2. Cycloheximide was reported to stabilize polysomes in rat liver (JONDORF et al., 1966) and reticulocytes (GODCHAUX et al., 1967), and to cause increased stability and storage of arginase mRNA in

of *P. cyclopium* 48 h p.i. were transfered to emerged conditions and grown as described in Fig. 23B. At indicated time interval, 100 μg/ml CH were added to nutrient solution. Abscissa: Hours after transfer to emerged conditions. ^{3}H-phe incorporation (A): Values in control cultures (= 100%) decreased during culture period to about one-half of initial value 35 h after transfer. Open and full squares represent two independent experiments. Free amino acids (A): 100% = 2.7 μmol/cm² culture area calculated for leucine as reference standard. Rates of cyclopenin-cyclopenol formation (B): 10 = 10 μg/hour · cm² culture area. For measurement of β-galactosidase inducibility (C) disks of glucose cultures were kept 3 h on nutrient solution containing 7.5 mg/ml arabinose and afterwards were analyzed for β-galactosidase (ININGER and NOVER, 1975); 100% = 30 and 6 mU/cm² culture area 40 h and 100 h after transfer, respectively. For longterm induction of β-galactosidase (D) 35 h after transfer, glucose in nutrient solution was replaced by arabinose. Cyclopenin-cyclopenol formation proceeded essentially as in glucose cultures; β-galactosidase: 10 = 120 mU/cm² culture area

Aspergillus nidulans (CYBIS and WEGLENSKI, 1972). However, because no enhancement of β-galactosidase formation in partially synchronized cultures of *P. cyclopium* was found after cycloheximide action (Fig. 27D), there is no physiologic evidence for an increased polysome content or accumulation of mRNA in this case.

3. The cycloheximide-dependent protein deficiency triggers an increased activity and/or de novo formation of the translational apparatus after the drug has been removed. In fact, the incorporation of ^{3}H-phenylalanine into TCA-precipitable material (Fig. 27A) as well as the inducibility of β-galactosidase (Fig. 27C) in the cycloheximide cultures after a period of strong inhibition is greater than that in the control cultures. However, despite certain relationships there are also distinct differences between the dynamics of the individual processes. The good correlation between ^{3}H-phenylalanine incorporation and alkaloid formation rates 40-80 hours after transfer (Fig. 27A, B) is in opposition to the lack of secondary inhibition in ^{3}H-phenylalanine incorporation. Furthermore, a direct connection between superinduction and overshooting of ^{3}H-phenylalanine incorporation into proteins is doubtful because the related effect of induction of the model enzyme β-galactosidase is observed only when the rate of alkaloid formation has already passed to the secondary inhibition phase (80 h after transfer). It is interesting to note that both β-glactosidase *inducibility* in glucose cultures exposed for 3 h to arabinose (Fig. 27C) and β-galactosidase *activity* in arabinose cultures (Fig. 27D) increase following a 10-hour lag-phase after cycloheximide removal (cf. arrows in Figs. 27C, D). At this time, ^{3}H-phenylalanine incorporation into the proteins surpasses control levels.

4. The role of metabolically unstable regulatory RNAs and proteins, leading to inactivation (and possible degradation) of the corresponding mRNAs or enzyme proteins (TOMKINS et al., 1972; FRENCH and SMITH, 1975; Chap. C4; cf. however, KILLEWICH et al., 1975), has been discussed in connection with related superinduction phenomena caused by numerous antibiotics, including cycloheximide, on induced enzyme synthesis in practically all kinds of organisms. Though the experimental evidence in some systems is very persuasive (cf. e.g., Chap. C4), the existence of such regulatory molecules has not yet been shown directly. Application of this concept to the regulation of alkaloid formation in *P. cyclopium* leads to the following highly speculative explanation of the superinduction phenomenon: The increase of the rate of alkaloid formation in the idiophase is controlled in general by a regulatory protein interfering with the formation of the rate-limiting protein(s) of alkaloid biosynthesis in vivo. Cycloheximide incubation leads to a significant decrease of the cellular concentration of the regulatory protein and consequently after restoration of protein synthesis to an abnormally high rate of the de novo formation of the supposed rate-limiting protein(s).

Summarizing, there are a number of specific effects following administration of cycloheximide but none of them appears to be directly correlated to the oscillation of alkaloid formation rates. Further analysis of these phenomena necessitates in vitro determination of the as yet unknown rate-limiting protein(s) of alkaloid biosynthesis, a problem closely connected with investigations of the possible role of enzyme degradation in the dynamics of cyclopenin-cyclopenol formation in general and in particular after cycloheximide administration.

d) Glucose as Repressor of Idiophase Development

Glucose repression is a common phenomenon in industrial microbiology as shown, e.g., for the production of penicillin, gibberellic acid, chloramphenicol, streptomycin, siomycin, and actinomycin (DEMAIN, 1968, 1971, 1972; GRISEBACH, 1975). The effect can also be demonstrated for submerged cultures of *P. cyclopium* (SCHRÖDER, unpublished results) that never produce alkaloids or conidiospores if cultivated in the medium (NLI) used for emerged cultivation. If, however, glucose is replaced by sorbitol and mannitol both idiophase characteristics are expressed normally. More detailed investigations revealed that the effect of glucose is superimposed by the positive or negative influence of other nutrient components (Ca^{2+}, NO_3^-, SCHRÖDER, unpublished results). Furthermore, addition of glucose to emerged grown alkaloid producing disks of idiophase cultures after submerging immediately reduces the rate of alkaloid synthesis by more than 50%, an effect that cannot be due to the inhibition of enzyme synthesis.

The decisive influence of the nutrient composition on the expression of idiophase characteristics observed with the submerged cultivation of *P. cyclopium* cannot be demonstrated in emerged cultures; alkaloid and conidia formation begins and proceeds despite an excess of glucose, ammonia, and phosphate. But emerged cultures present a serious drawback with respect to the investigation of such nutrient effects. Because growth occurs at the surface of the nutrient solution, there is strong polarization of the mycelium. Nutrients and oxygen are distributed unequally between the different layers of the mycelium; it is therefore impossible to estimate the influence of the nutrients simply from the composition of the nutrient solution. In this respect it is of interest that glucose-caused repression of alkaloid formation also in emerged cultures can be detected if cylcopenin-cyclopenol excretion by the hyphae is refered to the protein content and not as usual to the culture area (see above). The production rate per unit protein is higher under conditions of low glucose concentrations, e.g., after replacement with diluted nutrient solution. This effect becomes more prominent if glucose in the nutrient solution is replaced by other carbohydrates. Thus, in the presence of lactose and glycerol instead of glucose, the alkaloid production rate was stimulated 40-80%, whereas the protein content and the rates of protein synthesis and conidia formation were decreased by 20-60% (Fig. 28).

Efforts to investigate the possible molecular mechanism of the action of glucose on secondary metabolism were basically influenced by studies of glucose repression of catabolic enzymes in bacteria, especially in *Escherichia coli* (catabolite repression). These experiments revealed that glucose acts on the cellular concentration of cyclic adenosine-3',5'-monophosphate (cAMP), which as a nonsubstrate-like effector of the catabolite-activating pro-

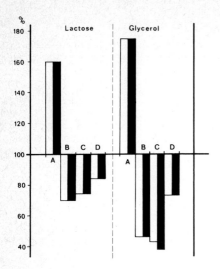

Fig. 28. Influence of carbohydrates and of cAMP on growth, alkaloid and conidia formation by *P. cyclopium* (ININGER, unpublished results). Emerged cultures were grown on nutrient solutions containing glucose (control), 40% glucose + 60% lactose, and 40% glucose + 60% glycerol, respectively, with or without cAMP (10^{-3} M final concentration). 48 h p.i. cultures were transferred to diluted nutrient solution containing corresponding carbon source: glucose (control), lactose, or glycerol. For determination of individual parameters, cf. NOVER and LUCKNER (1974). Values of control cultures on glucose nutrient solution are taken as 100%. □ - cAMP; ■ + cAMP; A: Rate of cyclopenin-cyclopenol production 168 h p.i. (control: 50 µg/mg protein/hour); B: rate of conidia formation 48-96 h p.i. (control: $2 \cdot 10^8$ spores/cm^2 culture area/48 h; C: protein content (control: 300 µg/cm^2 culture area, 72 h p.i.); D: relative rate of protein synthesis

tein is required for the transcription of the genes coding for many catabolic enzymes (PASTAN and PERLMAN, 1970). Employing intact cells, sugars were shown to have two primary effects on cAMP metabolism: (1) inhibition of synthesis while promoting degradation, and (2) stimulation of efflux of cAMP into the extracellular fluid (RICKENBERG, 1974; SAIER et al., 1975). By these effects addition of glucose rapidly lowers the endogenous level of cAMP, thereby causing repression of cAMP-dependent reactions.

In eukaryotic systems, the mechanism of glucose repression is unknown. In *P. cyclopium* the effect of glucose on alkaloid metabolism as well as on arabinose-induced β-galactosidase formation (ININGER and NOVER, 1975) could not be influenced in any experimental variant by administration of 10^{-3} M cAMP or dibutyryl cAMP. These results indicate that glucose repression in *P. cyclopium*, as, e.g., in yeast (HOLZER, 1975), is not cAMP-mediated. Recent experiments have shown that glucose inhibits β-galactosidase formation at one of the post-transcriptional levels (ININGER and NOVER, 1975).

e) The Developmental Program of Penicillium cyclopium

In Fig. 29 we have summarized the experimental results concerning the idiophase development of *P. cyclopium* discussed thus far from different points of view. The scheme is based on the temporary sequence of events during the mould's development (cf. Chap. D1a and D1b) and includes results of the influence of inhibitors of gene expression and DNA synthesis (cf. Chap. D1b and D1c), of the influence of glucose on idiophase events (cf. Chap. D1d), of mutants with characteristic deviations from the normal program of the wild-type strain as well as of the action of external factors (promotion of cyclopenin-cyclopenol synthesis by mycelial extracts and effectors of the phenylalanine type, see below). The scheme is directed toward the formation of the products of differential gene expression, i.e., the proteins that are thought to be representative of the idiophase development of *P. cyclopium*. However, it is an extreme simplification of the developmental processes and will have to be complemented and revised according to future experimental results. In particular, the possible causal relations between individual events and the regulatory mechanisms that participate in the overall process are almost completely unknown.

The following types of mutants listed in the scheme may be distinguished (SCHMIDT et al., Z. Allgem. Microbiol., in press):

1. Mutants in which all idiophase events are depressed: The defects of these strains presumably involve central regulatory events occurring at the beginning of the idiophase or in early trophophase, i.e., in a period described in Fig. 29 as determination phase. It is an interesting characteristic of one of these mutants (rev-meth 83a) that it spontaneously produces a considerable number of revertants, which closely resemble the wild-type strain in all properties.
2. Mutants in which only some idiophase events are affected: In strain dev 63, for instance, despite a normal alkaloid production rate in the hyphae, the amount of conidia is reduced to one tenth in favor of an increased mycelial dry weight. Furthermore, in the conidia there is no pigment synthesis and practically no cyclopenase formation. Strain aux-glu 1 has yellow-green conidiospores and decreased phenol oxidase activity as well as lowered pigment content in the conidia fraction. In mutant SM 72 b, which arises spontaneously in the wild-type strain, very little alkaloid is formed in the hyphae and conidia, whereas all other characteristics of the idiophase are normal. In another mutant, res-eth 1, besides its resistance to ethionine, the synthesis of the green pigment of the conidia is blocked. The block may be caused by a time-shift in the expression of phenol oxidase during the phase of conidiation (BARTSCH, unpublished). The properties of these mutants show that different segments of the idiophase may be deleted by mutation without serious inhibition of the overall process.

It is, however, evident that this kind of biochemical analysis of mutational blocks leads to rather circumstantial conceptions

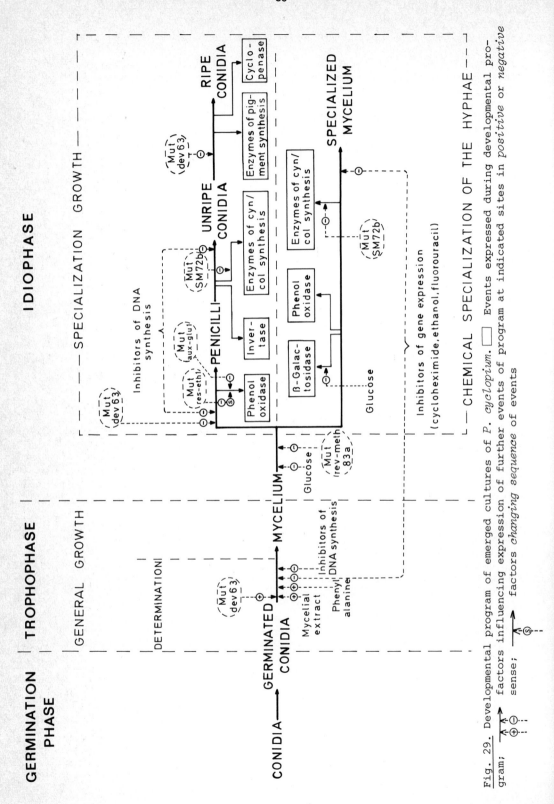

Fig. 29. Developmental program of emerged cultures of *P. cyclopium*. □ Events expressed during developmental program; ---○--- factors influencing expression of further events of program at indicated sites in *positive* or *negative* sense; ---Ⓢ--- factors *changing sequence* of events

of the kind of mutation involved. For instance, the following questions cannot at present be answered:

1. Is the defect in conidia ripening of mutant dev 63 a direct consequence of the changed quantitative relations between mycelial growth and conidia formation?
2. Are there any connections between ethionine resistance of res-eth 1 and its failure to produce the green pigment?
3. Is the low alkaloid production rate of mutant SM 72 b really due to the low expressibility of the corresponding genes or rather the consequence of changes in the precursor levels and/or stability of the enzymes involved in cyclopenin-cyclopenol biosynthesis?

There is no doubt that more detailed investigation of the procedure of individual steps of the overall specialization processes in these and other suitable mutants of *P. cyclopium* are necessary. Such investigations should of course also include genetic analyses of the mutational blocks.

The experimental evidence shown in Fig. 29 that led to the assumption of a distinct determinative period at early trophophase deserves special notice. The experiments showed that alkaloid production is stimulated by the precursor amino acid phenylalanine and by phenylalanine derivatives that are not incorporated into the alkaloids, if these compounds are administrated during the germination and early growth phases (DUNKEL et al., 1976; Fig. 30).

Furthermore, factors isolated from the mycelium of *P. cyclopium*, which are heat stable and have a molecular weight between 1000 and 5000, also increase alkaloid formation if added to the culture medium at the beginning of the growth phase, i.e., 0-8 h after inoculation. An early effect measurable in cultures treated with these "natural" effectors is pronounced acceleration of the growth phase. Protein content and rates of protein synthesis per unit culture area 48 h p.i. are about 2-3 times higher than in control cultures. Consecutively, sporulation and alkaloid production start earlier. The maximal production rates of cyclopenin and cyclopenol are increased by more than 100% compared with control cultures, though as early as the beginning of the idiophase (72 h p.i.) the described differences in protein synthesis and protein content between extract-treated and nontreated cultures have practically disappeared (Fig. 31). The increase of alkaloid production is paralleled by a higher level of activity of the enzymes involved, as shown for cyclopeptine dehydrogenase (EL KOUSY, unpublished results). Because the extracts used in these experiments do not contain any detectable amounts of amino acids, their activity certainly is not due to phenylalanine and phenylalanine derivatives.

The pronounced time interval between administration of the effector and the appearance of its action during the idiophase indicates that the stimulation of alkaloid biosynthesis is a secondary event, possibly evoked by enhanced formation of factors (regulatory proteins, endogenous effectors, etc.) that are required for the expression of the idiophase processes. However,

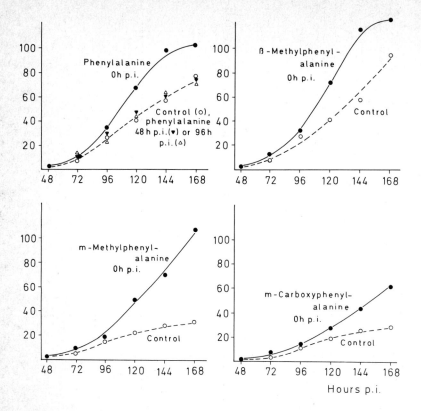

Fig. 30. Regulatory influence of phenylalanine and phenylalanine derivatives on cyclopenin-cyclopenol production rate in *P. cyclopium* (DUNKEL et al., 1976). Cultures were grown as described in Fig. 20. 3 µmol of D,L-phenylalanine or D,L-phenylalanine derivatives were added at indicated times for 48 h (if added at inoculation) or 24 h (if added 48 or 96 h after inoculation) Ordinate: 100 = 37 µg cyclopenin and cyclopenol/hour · cm^2 culture area

for a more detailed analysis of the chemistry and mode of action of the active principles in the mycelial extracts, mutants of *P. cyclopium*, which themselves are unable to produce the mycelial factors, are required in order to allow the reliable quantification and description of the metabolic events affected.

2. Biosynthesis of Secondary Products during Bacterial Sporulation

The formation of endospores by some bacteria, e.g., bacilli, is a complex sequence of morphologic and biochemical changes by the bacterial cell (HANSON et al., 1970; MANDELSTAM, 1976). Usually six stages of this differentiation program are distinguished (Table 4), whose expression involves at least 30 independently regulated gene groups with more than 100 genes scattered over the whole bacterial chromosome (HANSON et al., 1970; PIGGOT, 1973; COOTE and MANDELSTAM, 1973; MANDELSTAM, 1976). The mutual regulatory linkages of these genes were shown by mutants and

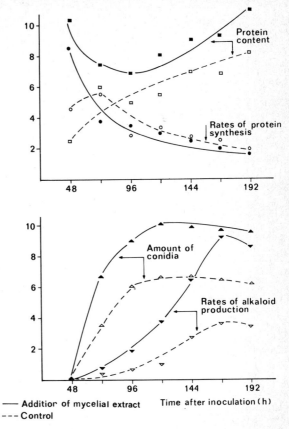

Fig. 31. Influence of mycelial extract on growth and specialization processes of emerged cultures of *P. cyclopium* (DUNKEL et al., 1966). Cultures were grown as described in Fig. 20. At time of inoculation, 1 ml extract (≙ 2 g fresh mycelium) was added to culture medium. Extract was prepared from 60 h-emerged grown mycelium of same strain of *P. cyclopium* by grinding with dry ice and extraction with acetic acid. After evaporation of acid *in vacuo* at 30°C, residue was dissolved in water and purified by chromatography on sephadex G 50. Protein content, protein synthesis, amount of conidia, and rate of cyclopenin and cyclopenol biosynthesis were determined as described by NOVER and LUCKNER (1974). Protein content (10 = 0.40 mg/cm² culture area); rates of protein synthesis (10 = 500 Imp./min); amount of conidia (10 = 3 · 10^8/cm² culture area); rates of cyclopenin and cyclopenol production (10 = 45 µg/cm² culture area)

use of inhibitors of gene expression. It was demonstrated that genetic blocks interfering with sporulation at specific stages prevent the expression of the cytologic and biochemical characteristics of the subsequent stages (WAITES et al., 1970; WOOD, 1971), i.e., sporulation is a highly coordinated sequence of closely coupled events (MANDELSTAM, 1976). DI CIOCCIO and STRAUSS (1973) demonstrated stage specific transcription patterns during sporulation. Additionally, experiments with actinomycin have shown that transcription and translation patterns during sporulation are not coincident and hence suggest a specific control of gene expression also at the translation level (STERLINI and MANDELSTAM, 1969).

Table 4. Morphologic and biochemical events associated with sporulation in *Bacillus* spec. (from MANDELSTAM, 1976; GREENLEAF et al., 1973; LEE et al., 1975; redrawn)

Stage	0	I	II	III	IV	V	VI
	0 h	1.5 h	2.5 h	4.5 h	6 h	7 h	8 h
Morphologic characteristics	Vegetative cell	Chromatin filament	Spore septum	Spore protoplast	Cortex formation (refractility)	Coat formation	Maturation
Biochemical characteristics		Exo-protease	Alanine dehydrogenase	70 000 dalton RNA binding protein	Ribosidase	Cysteine incorporation	Alanine racemase
			β-alanine and pantothenic acid uptake	Alkaline phosphatase	Adenosine deaminase	Octanol resistance	Heat resistance
			Peptide antibiotics	Glucose dehydrogenase	Sulfo-lactic acid	Brown pigment	
				Heat-resistant catalase	Dipicolinic acid		

At least 4 types of secondary products are formed during sporulation: peptide antibiotics during stage II (LEE et al., 1975), dipicolinic acid as well as sulfolactic acid during stage IV, and a brown pigment, whose structure is still unknown, during stage V (cf. Table 4). The kinetics of formation of tyrothricin, a complex of peptide antibiotics, of sulfolactic acid, and dipicolinic acid together with other biochemical events are shown in Figs. 32 and 33, respectively. As spore encapsulation progresses,

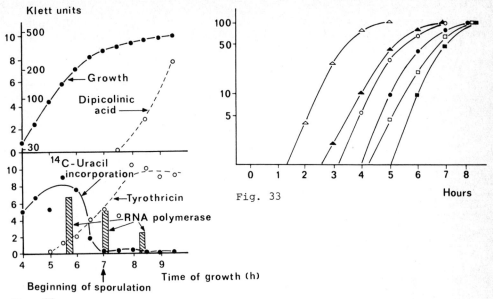

Fig. 32

Fig. 32. Time course of tyrothricin formation and other events during sporulation of *Bacillus brevis* (from SARKAR and PAULUS, 1972, redrawn). Growth (culture density, Klett units); dipicolinic acid: 10 = 2 µg/mg of cells; tyrothricin: 10 = 15 µg/ml of cells; ^{14}C-uracil incorporation into RNA: 10 = 150 cpm/mg of cells · min; RNA polymerase activity: 10 = 0.3 nmol ^{3}H-UMP-mg protein

Fig. 33. Time course of biochemical events during spore formation of *B. subtilis* (from WOOD, 1971). △—△ Alkaline phosphatase (100 = 14.2 U/mg protein); ▲—▲ glucose dehydrogenase (100 = 28.4 U/mg protein); ○—○ sulfolactic acid (arbitrary units corresponding to cpm after feeding $[^{35}S]$-sulfate); ●—● refractility (100 = 80% refractile spores); □—□ 2,6-dipicolinic acid (100 = 28 µg/mg protein); ■—■ heat resistance (100 = 80% heat-resistant spores)

the formation of peptide antibiotics migrates from the soluble fraction of the bacterial cell into the forespore, terminating with the separation of forespores from the sporangium membrane (LEE et al., 1975). Experiments with inhibitors of gene expression have shown that synthesis of the peptides depends on the de novo formation of RNA and protein (KURAHASHI, 1974).

The close integration of the formation of secondary products into the overall sporulation program is demonstrated by the behavior of mutants blocked at one of the sporulation stages (Table 5). The results justify the assumption that there is a sequence of dependent events and that in the wild-type strain the enzymatic changes listed in Table 4 are interdigitated with morphologic changes and the formation of the secondary products. This does not mean, however, that all events lie in the main line of the program. In fact, experimental evidence indicates that secondary product formation can be blocked without impairing the expression of the subsequent steps of sporulation. Despite many past speculations (WEINBERG, 1970; SARKAR and PAULUS, 1972; PAULUS and SARKAR, 1976), it appears that none of the secondary products has an es-

Table 5. Stage specific block of sporulation events in mutants of *B. subtilis* (from MANDELSTAM, 1976)

Strains	Wild-type	E22	T20	N25	NG 13	E33
Stage blocked	none	0	I	II	III	IV
Proteases	+	−	+	+	+	+
Peptide antibiotics	+	−	−	+	+	+
Alkaline phosphatase	+	−	−	−	+	+
Glucose dehydrogenase	+	−	−	−	−	+
Refractility	+	−	−	−	−	±[a]
Dipicolinic acid	+	−	−	−	−	−
Heat resistance	+	−	−	−	−	−

[a] Refractility feeble

sential function within the sporulation program (cf. KAMBE et al., 1974; HAAVIK and FRØYSHOV, 1975; and the review of MANDELSTAM, 1976).

The onset and procedure of sporulation is subjected to a quasi-catabolite repression. The sporulation program can be triggered at certain stages of the cell division cycle (MANDELSTAM, 1976) by glucose or nitrogen starvation, and, up to stage IV, it can be reversed by enrichment of the culture medium.

One of the earliest events is the formation of highly phosphorylated substances such as adenosine-5'-triphosphate-3'(2')-triphosphate (pppAppp) at the surface of the cell membrane. It is thought that these substances cause changes in the metabolism of the sporulating cells by several as yet unknown steps (RHAESE and GROSCURTH, 1976).

It was frequently argued that the onset of sporulation requires a basic change in the gene expression programs of bacteria evoked by the peptide antibiotics. In fact the peptide antibiotics were found to bind to bacterial DNA in vitro as well as in vivo and to influence RNA synthesis (SARKAR and PAULUS, 1972; SCHAZSCHNEIDER et al., 1974; RISTOW et al., 1975a-c; PAULUS and SARKAR, 1976). Other peptides, e.g., linear gramicidin, at least partly reverse the tyrocidine inhibition by a mechanism still unknown (RISTOW et al., 1975a, b). However, two types of results oppose a decisive role of the antibiotics for the sporulation process. On the one hand, peptides may be formed during the growth phase without triggering sporulation. On the other hand, some mutants incapable of producing antibiotics sporulate normally (KAMBE et al., 1974; HAAVIK and FRØYSHOV, 1975). Furthermore, proteolytic modification of RNA polymerase was suggested to cause the switch from transcription of the vegetative genes to that of the sporulation genes (PINE, 1972; LEIGHTON et al., 1973; KLIER et al., 1973) and indeed sporulating bacteria contain proteins which change the transcriptional specificity of the RNA polymerase probably by interference with the binding of σ-subunit to core enzyme (LOSICK and PERO, 1976). However, thorough experiments performed recently indicate that all *proteolytic* modifications so far reported are in vitro artifacts due to the increased formation of proteolytic enzymes during sporulation (LINN et al., 1973; REXER et al., 1975).

3. Sequential Gene Expression in Secondary Metabolism of *Penicillium urticae*?

The subsequent formation of enzymes of metabolic pathways by sequential gene expression is a well-known phenomenon in the catabolic metabolism of bacteria (cf. ORNSTON, 1971). In mandelic acid degradation by *Pseudomonas* species, for instance, several groups of enzymes form a pathway, which from mandelic acid leads finally to succinate and acetyl CoA. The subsequent steps of the pathway are induced by the sequential action of D-mandelic acid and several intermediates of its degradation (benzoic acid, cis, cis-muconate, and β-ketoadipate). The inductive actions of these compounds are superimposed by multiple end product repression and catabolite repression, integrating the pathway into a network of metabolic regulations.

Though in secondary metabolism there are several examples suggesting the sequential expression of genes involved in a particular pathway, the experimental data supporting this concept are scarce. In the following, the regulation of patulin biosynthesis in *P. urticae*, which has been investigated by several groups for many years, is discussed in more detail.

Patulin, a polyketide antibiotic, derives from acetyl and malonyl CoA via 6-methyl salicylic acid (6-MSA) and the intermediates shown in Fig. 34. BU'LOCK et al. (1965) demonstrated that four periods can be distinguished for patulin biosynthesis during the development of batch cultures of *P. urticae* (Fig. 35): (1) Trophophase - No formation of 6-MSA or of the other secondary products occurs. 6-MSA added to the cultures is not metabolized. (2) Early idiophase - The onset of 6-MSA synthesis is indicated by its accumulation in the culture medium. (3) Middle idiophase - 6-MSA concentration reaches a maximum and than drops to a steady state level. Concomitantly the accumulation of gentisyl derivatives begins. (4) Late idiophase - The biosynthesis of patulin starts after the gentisyl derivatives reach a certain concentration in the culture medium.

From the sequential appearance of 6-MSA, the gentisyl derivatives, and patulin as well as from the observation that 6-MSA, if applied externally to early idiophase cultures, evokes premature conversion to gentisyl derivatives, a sequential expression of the genes involved in patulin biosynthesis by intermediates of the pathway was suggested (BU'LOCK et al., 1965). However, a detailed analysis

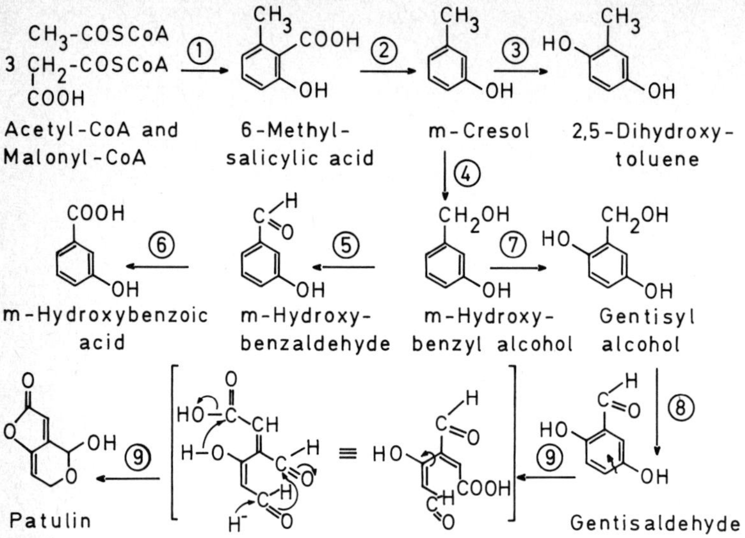

Fig. 34. Pathway of patulin biosynthesis in P. urticae (from MURPHY and LYNEN, 1975, redrawn). (1) 6-MSA synthetase complex; (2) 6-MSA decarboxylase; (3) m-cresol 6-hydroxylase; (4) m-cresol methylhydroxylase; (5) m-hydroxy-benzyl alcohol dehydrogenase; (6) m-hydroxybenzaldehyde dehydrogenase; (7) m-hydroxybenzyl alcohol 6-hydroxylase; (8) gentisyl alcohol dehydrogenase; (9) patulin synthetase

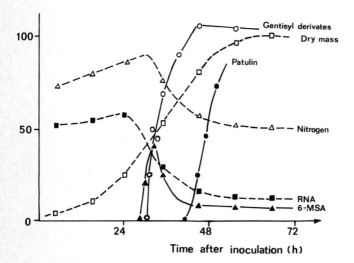

Fig. 35. Biochemical events during development of batch cultures of P. urticae (from BU'LOCK et al., 1965, redrawn). Cultures were grown on Czapek-Dox medium on rotatory shaker. Dry mass of mycelium (mg/30 ml); RNA content of mycelium (μg/mg dry mass); nitrogen content of mycelium (μg/mg dry mass); 6-MSA (100 = 5 mmol); gentisyl derivatives (100 = 1 mmol); patulin (100 = 0.2 mmol)

of the regulatory aspects revealed that the picture is more complicated.

The expression of the complete pathway is evidently controlled by the nutrient conditions (BU'LOCK et al., 1965, 1969; LIGHT, 1969; FORRESTER and GAUCHER, 1972). No secondary products are formed by the cultures on nutrient-rich media. Possibly cAMP takes part in this kind of catabolite repression because m-hydroxybenzyl alcohol dehydrogenase activity significantly increases after addition of dibutyryl cAMP under conditions of high nutrient supply (FORRESTER and GAUCHER, 1972).

Whereas inhibitors of gene expression (e.g., cycloheximide) in some strains prevent the appearance of 6-MSA synthetase activity if added at the beginning of the production phase (LIGHT, 1967, 1970), in others addition of p-fluorophenylalanine, β-thienylalanine, or cycloheximide in the first stage of the idiophase barely affects the rate of 6-MSA production and only inhibits 6-MSA conversion (BU'LOCK et al., 1969). This result indicates that 6-MSA synthetase in these strains is formed before 6-MSA synthesis really starts. A similar discrepancy between enzyme and product formation was found in emerged cultures of *P. cyclopium* under some culture conditions (cf. Chap. D1a). However, so far there are only speculations about the molecular basis of these effects.

The in vitro determination of 6-MSA synthetase and 6-MSA decarboxylase (Fig. 34, Nos. 1 and 2) demonstrates coincident expression in cultures of an early strain immediately after resuspension (LIGHT, 1969). On the other hand benzyl alcohol dehydrogenase (Fig. 34, No. 5), which is not involved in the patulin pathway, in fermentor cultures is biphasic, exhibiting a second maximum following that of 6-MSA concentration in the medium. Its activity thus seems to be independently regulated.

In summary, evidence for sequential induction of the enzymes of patulin formation by intermediates of the pathway only comes from the inhibitor experiments of BU'LOCK et al. (1969). An influence of 6-MSA or other intermediates of the biosynthetic chain on the formation of the consecutive enzymes could not be substantiated in recent experiments (MURPHY and LYNEN, 1975).

4. Sequential Formation of Secondary Products during the Generative Development of Mucoraceous Fungi

The sexual reproduction of heterothallic species of the family of Mucoraceae is triggered if two mycelia of different mating types, designated (+) and (-), come close to each other. It involves a complex sequence of morphologic events, the initial stages of which are the following: (1) Hyphal growth is specialized to form so-called zygophores, i.e., sexually differentiated hyphae. The zygophores are homologous to another type of specialized hyphae, the asexual sporangiophores. Under certain conditions, transspecialization is observed between both types. (2) The zygophores of opposite sex come into contact with each other either by chance or by chemotropism (zygotropism), due to volatile "sex hormones" excreted by the hyphae of the opposite mating type.

After the contact between the zygophores, a series of further morphologic changes is initiated, which finally lead to the formation of gametangiae and zygospores. The latter are large, resistent cells with a thick, hardened, dark wall (for a detailed description and references to the literature cf. BURGEFF, 1924 and SUTTER, 1975).

Despite differences in numerous details, this generative idiophase development is basically similar in many species of the Mucoraceae, e.g., in *Mucor mucedo*, *Phycomyces blakesleeanus*, and *Blakeslea trispora*. It is closely connected with the sequential formation of secondary products derived from carotenes, which have an important function for the overall process.

More than 50 years ago BURGEFF (1924) observed that diffusible and/or volatile substances trigger zygophore formation and zygotropism. The supposed "sex hormone" was thought later to be a mixture of C_{13}-terpenoids, the so-called trisporic acids (Fig. 36, cf. BU'LOCK et al., 1972, and the summary of GOODAY, 1974).

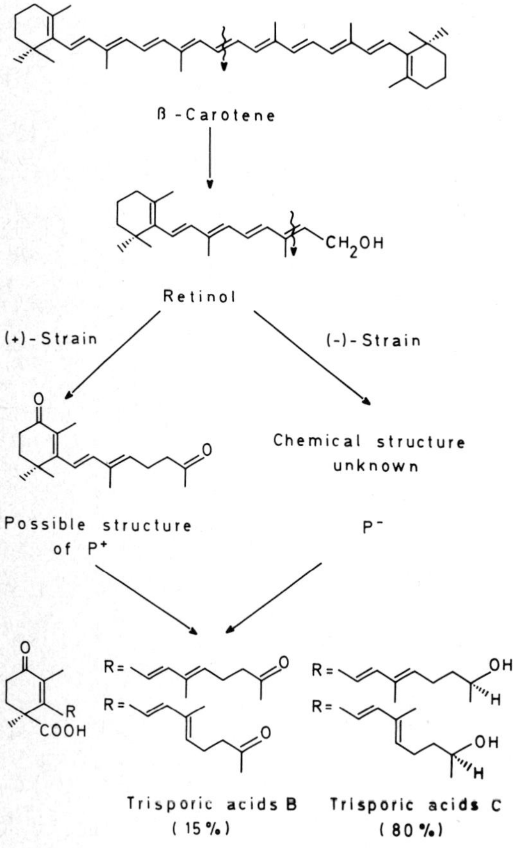

Fig. 36. Possible biosynthetic pathway to main constituents of trisporic acid group (P^+ and P^- are presumptive sex specific precursors)

Trisporic acids (TA) are formed from β-carotene in mature (+)- and (-)-mycelia if both come close to each other. The biosynthesis needs the constant exchange between the mycelia of diffusible products, which are probably precursors of TA (VAN DEN ENDE et al., 1972; GOODAY, 1974; SUTTER, 1975).

There is evidence that the onset of TA synthesis by mutual conversion of the sex-specific precursors by competent (+)-and (-)-mycelia, respectively, does not require RNA and protein synthesis, because a low level of the corresponding enzymes is present in the idiophase hyphae. However, no TA formation occurs if either of the mycelia is pretreated with 5-fluorouracil, a result indicating a true complementary synthesis. In contrast to the low basal level of TA-forming enzymes in idiophase mycelia, there is evidently a considerable increase of the enzyme amount after the first molecules of TA have been synthesized. This autoinduction is also inhibited by fluorouracil (WERKMAN and VAN DEN ENDE, 1973).

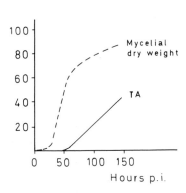

Fig. 37

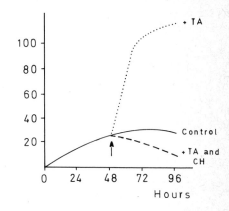

Fig. 38

Fig. 37. Growth and production of trisporic acids in mated submerged cultures of B. trispora (from VAN DEN ENDE et al., 1970). (+)- and (-)-strains of B. trispora were cultivated together under submerged conditions. Trisporic acids (TA) were extracted from culture medium with ether and after evaporation of solvent were determined polarographically. Amount is given as relative polarographic activity. Mycelial dry weight: 100 = 0.5 g/100 ml culture

Fig. 38. Stimulation of carotenogenesis in B. trispora (-)-strains by trisporic acid (from THOMAS et al., 1967). Submerged cultures of B. trispora 48 h p.i. were resuspended in fresh medium; 48 h later an extract containing TA and cycloheximide (50 µg/ml), respectively, was added. Ordinate: µg carotene/ 100 ml culture

Similar behavior is observed if (+)- and (-)-strains are grown together under submerged conditions (Fig. 37). After a trophophase of about 40 hours the rapid hyphal growth stops and large amounts of TA are excreted into the culture medium as typical idiophase products. Addition of inhibitors of gene expression (5-fluorouracil, cycloheximide) at any time of the accumulation period results in a rapid cessation of TA formation indicating the constant requirement of RNA and protein synthesis (cf. VAN DEN ENDE et al., 1970).

The morphogenetic action of TA, i.e., zygophore formation, is observed only in emerged cultures. All four compounds (Fig. 36) are almost equally active on (+)- and (-)-mycelia (BU'LOCK et al., 1972). Simultaneously with the outgrowth of the sexually differentiated hyphae there is a chemical specialization indicated by the accumulation of large amounts of carotenes, especially β-carotene. This process can be triggered also in submerged cultures of B. trispora (-)-mycelia. It is presumably part of the autoinduction phenomenon of TA described previously and is sharply inhibited by adding cycloheximide (Fig. 38). The excess carotene may have a dual function (GOODAY, 1974): (1) It serves as precursor for the sex hormone synthesis. (2) It is transformed into an oxygenated polymere, sporopollenin, which makes about 50% of the zygospore wall (SHAW, 1971; cf. also Chap. D6).

The understanding of the biochemical role of TA has recently made major advances through the following observations:

1. Large amounts of trisporic acids B and C are required to stimulate zygophore formation in Phycomyces blakesleeanus, indicating that the trisporic acids either are only metabolites of the supposed sex hormone or do not penetrate the hyphae of some species (SUTTER, 1975).
2. In accordance with the early observations of BURGEFF (1924), volatile, neutral precursors of trisporic acids were extracted from Mucor mucedo and shown to induce sex-specific zygophore formation (MESLAND et al., 1974). In contrast TA never evokes zygophore formation in one mating type only. These volatile, actual sex hormones, the chemical structures of which are only partially known (cf. Fig. 36), are found in higher concentrations in mycelia stimulated by TA (WERKMAN and VAN DEN ENDE, 1973). They are formed also in mature vegetative mycelia and may be involved in the chemotropic reactions of the zygophores (BURGEFF, 1924; MESLAND et al., 1974).
3. A considerable number of mutants of carotene metabolism were isolated from wild-type strains of P. blakesleeanus and tested for their mutual complementation in the triggering of zygophore formation and development (SUTTER, 1975). The results support the conclusion that TA are formed from β-carotene by the exchange of mating type-specific precursors. But they are intrahyphal regulators rather than interhyphal chemical messengers, i.e., the sex hormone(s).

In summary, during the transition from mature vegetative hyphae to zygospore formation the following reactions of isoprenoid metabolism are sequentially expressed (Fig. 39): (1) At the beginning of the early idiophase small amounts of sex-specific precursors of TA (P^+ and P^-) are synthesized and released into the surrounding milieu. (2) These precursors are transformed into TA by the hyphae of the opposite mating type. (3) The TA formed induces increased synthesis of β-carotene, i.e., its own precursor. This induction process is dependent on RNA and protein synthesis. (4) TA and/or the sex-specific precursors evoke the outgrowth of zygophores. The zygotropism observed in some species (Mucor spec.) may also depend on the sex-specific precursors. (5) As the last step, the oxidative polymerization of carotenes to sporopollenin is part of the differentiation program "zygospore ripening."

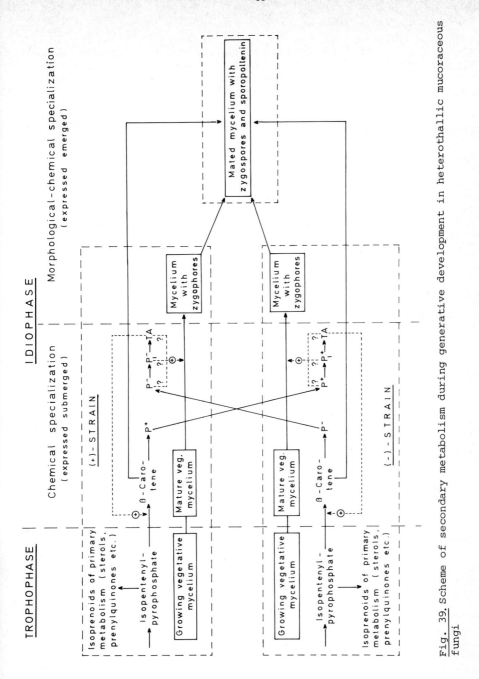

Fig. 39. Scheme of secondary metabolism during generative development in heterothallic mucoraceous fungi

The sequential formation of secondary products as an integrated part of the generative development of mucoraceous fungi is an excellent example of the connections between chemical and morphological specialization. The isolation of mutants with a block in carotene metabolism and as a consequence with an impaired sexual development (SUTTER, 1975) offers the basis for a more detailed analysis of the regulatory principles involved in this complex differentiation program, including also investigations on the enzymatics of carotene and TA biosynthesis.

5. Expression of Secondary Metabolism in Developing Chloroplasts

The biosynthesis of chloroplasts proceeds in a complex differentiation program comprising the de novo formation of specific tRNA and rRNA species as well as that of many proteins (aminoacyl-tRNA synthetases, ribosomal proteins, electron carriers, structural proteins of the membranes, enzymes participating in the reductive CO_2 fixation, etc.; PARTHIER et al., 1975). Most of these new components are coded by the nuclear DNA, but information for some of them is also stored in the DNA of the chloroplasts. The formation of photosynthetically active chloroplasts can be readily studied by illuminating bleached *Euglena* cells or etiolated seedlings of higher plants. Both systems contain precursors of the mature chloroplasts, so-called proplastids and etioplasts, respectively. At the onset of exposure to light, a complex, highly coordinated series of differential gene expression is initiated, which requires constant regulatory interaction between the two information systems involved. However, the details of the procedure and above all, the regulatory principles are still almost completely unknown (SCHIFF, 1974; PARTHIER et al., 1975).

There is increasing evidence that some types of secondary products are formed in chloroplasts and that the biosynthesis of the corresponding enzymes is regulated by light. In developing barley chloroplasts the formation of 6-methylsalicylic acid (6-MSA, Fig. 34) was observed (KANNANGARA et al., 1971). Its biosynthesis from acetate with isolated chloroplasts is dependent on the presence of coenzyme A and CO_2. As shown in Table 6, 6-MSA formation is negligible in etioplasts or mature chloroplasts and restricted to a short period of etioplast maturation where the active synthesis of photosynthetic membranes and grana proceeds. Unfortunately there are no studies to show whether an enzyme similar to the 6-MSA synthetase of *P. urticae* (cf. Chap. D3) is formed during the etioplast maturation or what other enzymes may be involved.

The frequent occurrence of flavonoids and other products of phenylpropanoid metabolism in chloroplasts (for a summary of the literature cf. SAUNDERS and McCLURE, 1975) suggests that their biosynthesis may also proceed within these organelles. Indeed the key enzyme of the pathway, phenylalanine ammonia lyase (Fig. 1, No. 1), was detected in chloroplasts of a considerable number of plant species (LÖFFELHARDT et al., 1973). About 17% and 29% of the total cellular PAL content was found in the chloroplast fraction of the green alga *Dunaliella marina* (LÖFFELHARDT et al., 1973) and of barley leaves, respectively (SAUNDERS and McCLURE, 1975). PAL content was highest in plastids of greening leaves. Light stimulation is evidently mediated by the phytochrome system (SAUNDERS and McCLURE, 1975; SCHOPFER et al., 1975; THIEN and SCHOPFER, 1975, cf. Chap. C4).

Table 6. Incorporation of acetate-$[1-^{14}C]$ into 6-MSA in plastids of barley leaves at different stages of greening (from KANNANGARA et al., 1971)

Illumination (Hours)	Chlorophyll content of chloroplast suspensions (mg/ml)	^{14}C-Incorporation into ether-soluble material (cpm)	Proportion of ^{14}C in 6-MSA (%)
0	0.001	32 280	2.1
6	0.367	60 920	39.1
24	2.815	116 208	0.0

Plastids were isolated at time points indicated in table. Purified plastid fraction was incubated in vitro for 45 min at 3000 lux and 25°C. Incorporation of radioactivity from acetate into 6-MSA was determined after thin-layer chromatographic purification and recrystallization.

Among the products of phenylpropanoid metabolism of chloroplasts, the benzoic acid derivatives have been investigated in some detail (CZICHI and KINDL, 1975; LÖFFELHARDT and KINDL, 1975). The enzymes catalyzing the synthesis of benzoic acid and p-hydroxybenzoic acid from phenylalanine are bound to the thylakoid membrane. The cinnamic acid formed as an intermediate is evidently channelled to the subsequent enzyme without exchanging with the pool of free cinnamic acid. For the further transformation of cinnamic acid or its p-hydroxy derivatives, a "benzoate synthetase complex" is postulated that might function analogously to the fatty acid degrading enzyme system (CZICHI and KINDL, 1975; LÖFFELHARDT and KINDL, 1975).

In summary, though studies of the expression of secondary metabolism in chloroplasts are just beginning, a number of important features have already evolved: (1) There is a considerable amount of PAL and perhaps also of the other enzymes of cinnamic metabolism (cf. Fig. 1) localized in the thylakoid membranes. These enzymes may be regulated independently from the corresponding enzymes (isoenzymes?) in other cell compartments. (2) The organization of the enzymes into a membrane-bound complex is combined with strict separation of some of the intermediates from their soluble pools. (3) The light-dependent greening of etioplasts coincides with a relatively pronounced activity increase of the enzymes of chloroplast secondary metabolism. Moreover, formation of some of the enzymes seems to be restricted to the greening period.

These results afford the basis for an intensive investigation of the formation of some types of secondary metabolites in greening chloroplasts, of their possible role in the mature organelle, and of the regulation of the enzymes involved.

6. Secondary Product Formation during Microsporogenesis in Higher Plants

Microsporogenesis is a prominent feature of anther development. From the sporogene tissue microspore mother cells are formed that undergo meiotic cleavage to yield the spore tetrads. This part of microsporogenesis may be compared with the trophophase in cultures of microorganisms. The "idiophase" comprises microspore maturation, i.e., disintegration of the tetrads by degradation of the surrounding callose and stepwise formation of the typical pollen wall (HESLOP-HARRISON, 1971; STANLEY and LINSKENS, 1974).

The outer part of the pollen wall, the exine, is characterized by its large amount of sporopollenin, a secondary product derived by oxidative polymerization from carotenoids (BROOKS et al., 1971) and deposited as a population of radially directed columns, the so-called bacula. Just before the critical period of exine formation the carotenoid content of the anthers increases rapidly together with the activity of peroxidase, an enzyme that seems to be involved in sporopollenin synthesis (SHAW, 1971; cf. Fig. 40).

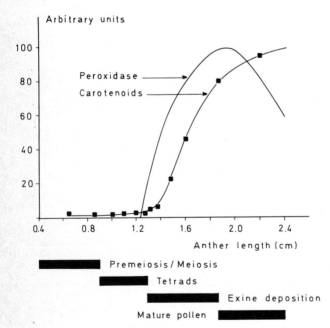

Fig. 40. Carotenoid content and peroxidase activity in developing anthers of *Lilium henryi* (from SHAW, 1971, redrawn)

The complicated spatial orientation of sporopollenin deposition is primed by a morphogenetic pattern of the endoplasmic reticulum at the plasmalemma of the pollen cells, which leads to a preferential binding of sporopollenin precursors to distinct sites of the unsculptured inner layer of the exine (HESLOP-HARRISON, 1971). Sporopollenin and peroxidase are produced by the tapetum cells surrounding the immature pollen grains. The final shape of the pollen exine thus is the result of a morphogenetic process requiring cooperation between the immature pollen grain and the tapetum cells. The pattern formation of exines provides a supreme example of the morphogenetic capacity of a single cell. There is evidence that the morphogenetic genes are transcribed in the pollen mother cells and that the meiotic cleavage has a major influence on the ordered deposition of sporopollenin during pollen maturation (HESLOP-HARRISON, 1971).

Another group of secondary products found during pollen maturation are the flavonoids (WIERMANN, 1973; QUAST and WIERMANN, 1973; SÜTFELD and WIERMANN, 1974; STANLEY and LINSKENS, 1974). Synthesis during microsporogenesis proceeds sequentially. A summary of the studies on anthers of *Tulipa* is given in Fig. 41. The compounds

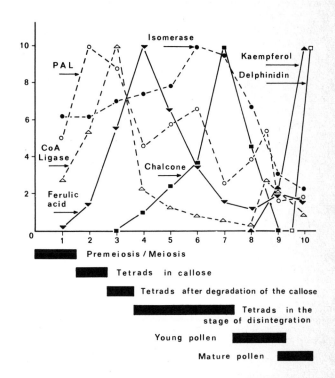

Fig. 41. Phase dependence of secondary product formation in anthers of *Tulipa* (from WIERMANN, 1973 and SÜTFELD and WIERMANN, 1974, redrawn). PAL (phenylalanine ammonia-lyase): 10 = 1.3 mU/mg protein; CoA ligase (p-coumarate: CoA ligase): 10 = 6.2 mU/mg protein; ferulic acid: 10 = 55 μmol/mg dry weight; isomerase (chalcone-flavanone isomerase): 10 = 2.0 ΔE_{366}/mg protein · min; chalcone (2', 3, 4, 4', 6'-pentahydroxychalcone): 10 = 6.2 nmol/mg dry weight; kaempferol: 10 = 4.0 nmol/mg dry weight; delphinidin: 10 = 50 nmol/mg dry weight

of phenylpropanoid metabolism are detectable in a sequence of four groups. The first group includes cinnamic acid derivatives (ferulic and p-coumaric acids) followed by chalcones, flavonol-type compounds (kaempferol, quercetin, isorhamnetin), and finally by delphinidin, a representative of the anthocyanins. This phase dependent formation of secondary metabolites was shown to be correlated with the dynamics of enzymes involved in their biosynthesis (cf. Fig 41). As expected, the peak activities of PAL and p-coumarate:CoA ligase (Fig. 1, Nos. 1 and 4) precede the maximum accumulation of cinnamic acid derivatives, and the peak activity of chalcone-flavanone isomerase (Fig. 2, No. 2) precedes in time the greatest chalcone concentration. It is therefore likely that the subsequent appearance of the different groups of secondary products is brought about by the sequential formation of the corresponding enzymes.

7. Induction of Urea Cycle Enzymes as Part of the Thyroid Hormone-Stimulated Metamorphosis of *Rana catesbeiana* Tadpoles

The formation of the secondary product urea (Fig. 42) in the liver cells of higher animals is a typical example of the excretory metabolism required to maintain the nitrogen balance of the organism. During the anuran development a shift from ammonotelism (excretion of ammonia) to ureotelism (excretion of urea) proceeds as a requirement of the transition from the aquatic life of the

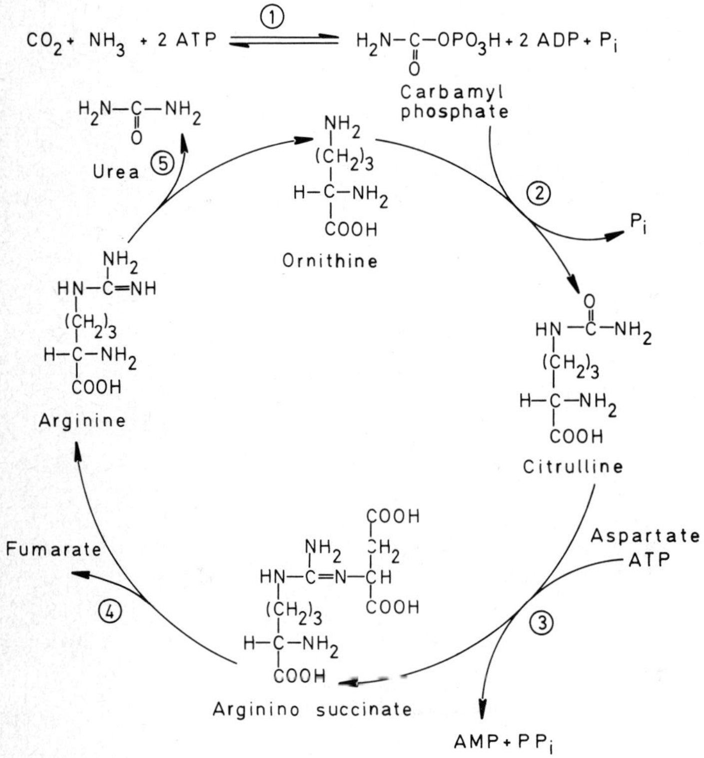

Fig. 42. The urea cycle. (1) Carbamyl phosphate synthetase, (2) orinithine transcarbamylase, (3) argininosuccinate synthetase, (4) argininosuccinase, (5) arginase

tadpole to the terrestrial life of the adult organism. The beginning of the ureotelism at stages IXX-XX of the normal anuran development is preceded by the de novo formation of the urea cycle enzymes (Fig. 43).

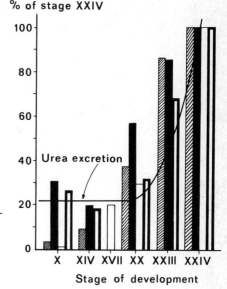

Fig. 43. Urea excretion and development of enzymes of urea cycle in metamorphosing tadpole of *R. catesbeiana* (from COHEN and BROWN, 1960, redrawn). ▨ Carbamyl phosphate synthetase; ■ ornithine transcarbamylase; ☐ arginine synthetase; ▥ arginase. Developmental stages indicated by roman letters are those defined by TAYLER and KOLLROS (Anat. Record 94, 7, 1946)

Table 7. De novo formation of proteins in tissues and organs of tadpoles during metamorphosis (FRIEDEN and JUST, 1970)

Liver	Enzymes of urea cycle, catalase, serum albumin, glutamate dehydrogenase
Tail	Hydrolytic enzymes: catepsin, phosphatase, RNase, DNase, β-glucuronidase
Red blood cells	Frog hemoglobin
Pancreas acinar cells	Acid phosphatase, α-amylase isoenzyme

The transition from ammono- to ureotelism is part of a sequence of complex changes in the metabolic capacity of the liver cells in this period (Fig. 43 and Table 7). The possibility to induce metamorphosis by external application of thyroid hormones 12-18 months before spontaneous metamorphosis has offered the experimental basis for a thorough analysis of this sequence. Long before the emergence of any new enzyme activity can be detected there is a pronounced increase of RNA synthesis (nuclear RNA, ribosomal RNA, transfer RNA), of ribosome and membrane formation, and of amino acid incorporation into proteins (Fig. 44; cf. the summaries by FRIEDEN and JUST, 1970 and TATA, 1970). Only after about five days do the characteristic new proteins, including the enzymes of the urea cycle, become detectable. This type of sequence is reminiscent of the metabolic changes in mammalian cells

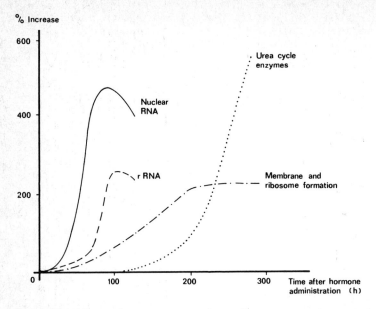

Fig. 44. Schematic representation of sequence of metabolic changes in liver of 3,3',5'-triiodothyronine-stimulated *R. catesbeiana* tadpoles (from TATA, 1970, redrawn)

found after estrogene administration to ovariectomized animals (HAMILTON, 1968). It indicates the principal difficulties encountered with such experimental systems. Though it is possible to trigger the premature expression of genes by exogenous addition of the natural effector, the formation of the particular proteins is preceded by a considerable number of more general effects. Their biosynthesis is only a late step of a differentiation program the kind and procedure of which is determined by the regulatory state of the individual cells. This is especially prominent when considering the multiplicity of metabolic changes evoked by thyroid hormones in different tissues of the tadpole (cf. Table 7 and the summary by FRIEDEN and JUST, 1970).

These results demonstrate that the expression of the capacity to form the secondary product urea in liver cells of anuran larvae during metamorphosis is an integral part of a complex developmental program of the whole organism. However, it remains to be shown whether the thyroid hormone administered to the intact animal is really responsible for the multiple effects in the different tissues or whether it is only the first link in an as yet unknown signal chain. Experiments in which thyroid hormones are administered directly are necessary to show the binding capacity and competence of the individual tissues of the tadpole for this hormone.

8. The Formation of Tanning Agents during Ecdysone-Controlled Pupation of *Calliphora* Larvae

The development of the blowfly, *Calliphora erythrocephala*, offers another example of secondary product formation as part of complex differentiation programs of animals, which are triggered by hormone action. The development of the blowfly from the egg to the pupa takes about 8 days and involves 3 larval instars. The moltings in between are initiated by the simultaneous secretion of juvenile hormone (a sesquiterpene) and ecdysone (a steroid), which are produced in the corpora allata and the prothoracic glands, respectively. If at the end of the third larval stage the production of juvenile hormone ceases, the following molting, which proceeds under the influence of ecdysone alone, leads to the pupa (KARLSON and SCHWEIGER, 1961; SEKERIS, 1965).

A target tissue of ecdysone during pupation is the epidermis of the late third instar larva and the white prepupa, respectively. Tyrosine degradation via p-hydroxyphenyl pyruvic acid in the early third instar larva is changed to the formation of dopa, dopamine, and N-acetyl dopamine. For the tanning of the cuticle of the pupa these new tyrosine metabolites play a dual role (CORRIGAN, 1970): (1) in the sclerotization by incorporation of oxidation products of N-acetyl dopamin into the protein matrix (fixation; KARLSON et al., 1969) and (2) in the melanization as a result of the accumulation of dark pigments (melanin) derived from dopa.

The broad investigations of the Karlson group concerning the regulation of pupation and especially the action of ecdysone on the larval epidermis have produced the tentative scheme of events shown in Fig. 45 (KARLSON, 1965; SEKERIS, 1965, 1969). The key enzyme of the new branch of tyrosine metabolism, dopa decarboxylase, is formed immediately after the onset of ecdysone excretion by the prothoracic gland and has maximal activity in the prepupa (Fig. 46). In larvae in which the abdomen was separated from the ecdysone source by a ligature between the third and fourth segments, dopa decarboxylase and the formation of the normal pupal cuticle is prevented but can be induced by injections of ecdysone (KARLSON and SEKERIS, 1962). The de novo synthesis of dopa decarboxylase and the pupal molting of normally developing third instar larvae is inhibited by antibiotics affecting RNA and protein synthesis (actinomycin, puromycin, streptomycin). The inhibitory effects have a pronounced optimum if the antibiotics are added 20-30 h before molting (SEKERIS and KARLSON, 1964; SEKERIS, 1965).

The increased de novo synthesis of dopa decarboxylase after ecdysone action was directly demonstrated also by double labeling and by immunoprecipitation of the enzyme. The amount of radioactively labeled dopa decarboxylase increases parallel to the ecdysone titer of the larvae. Since there was no significant difference in the half-life of dopa decarboxylase in noninduced and ecdysone-induced animals, the increased amount of the enzyme must be due to accelerated de novo synthesis (FRAGOULIS and SEKERIS, 1975a).

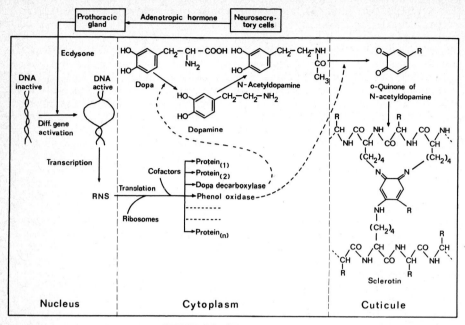

Fig. 45. Scheme of ecdysone action on differential gene expression in epidermis cells of *Calliphora* larvae; synthesis of tanning quinones from dopa (from SEKERIS, 1965, 1969, redrawn)

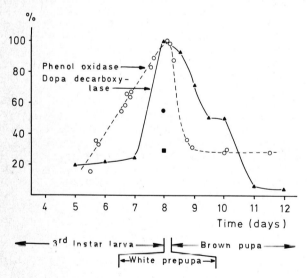

Fig. 46. Relative activities of dopa decarboxylase and phenol oxidase during embryonal development of *C. erythrocephala* (from KARLSON and SCHWEIGER, 1961 and FRAGOULIS and SEKERIS, 1975a, redrawn). ▲ Dopa decarboxylase activity in normal developing larvae; ■ enzyme activity in ligated 7-day-old larvae, 24 h after ligation; ● enzyme activity in ligated 7-day-old larvae, 24 h after ligation, which received 0.1 μg ecdysone 6 h before death; ○ phenol oxidase activity in normally developing larvae

Another enzyme of tyrosine metabolism influenced in its activity by ecdysone is the phenol oxidase catalyzing the formation of the o-quinone derivative from N-acetyl dopamine in the cuticle (Fig. 46). An inactive precursor of phenol oxidase is present in the hemolymph of the larvae (for a summary of the older literature cf. ASHIDA and ONISHI, 1967). Under the action of ecdysone in the cuticle cells a protein is produced (set free?) that activates the proenzyme (SEKERIS and MERGENHAGEN, 1964; KARLSON et al., 1964). Because it was found that proteolytic enzymes, e.g., α-chymotrypsin and aminopeptidase, to a certain degree are able to activate the enzymogen (SCHWEIGER and KARLSON, 1962), it was suggested that the activating protein likewise acts by restricted proteolysis (proteinogen processing, cf. Chap. A).

The detailed mode of action of ecdysone has been partially elucidated. Injection of the hormone into late instar larvae of *Chironomus tentans* or other related insects is followed shortly after by puffing of two particular loci in the giant salivary gland chromosomes (CLEVER and KARLSON, 1960; CLEVER, 1965). This puffing phenomenon was shown to be connnected with an intensive de novo synthesis of RNA (BEERMANN, 1972). In the epidermis of *Calliphora* larvae the nuclei bind ecdysone, and one of the first reactions to be noticed in vitro is an increase of RNA synthesis. Later, the RNA is bound in ribonucleoprotein particles, found in the polysomal fractions. Within 3 hours after ecdysone administration, the formation of polysomes significantly increases (MARMARAS and SEKERIS, 1972). The formation of new RNA isolated from the polysomal fractions is inhibited by α-amanitin acting on the initiation of transcription by the DNA-dependent RNA polymerase (SHAAYA and SEKERIS, 1971). α-Amanitin also inhibits the de novo synthesis of dopa decarboxylase.

By in vitro translation in a reconstituted system, it was demonstrated that mRNA coding for dopa decarboxylase is present in the RNA fraction of *Calliphora*. The mRNA was enriched by oligo(T)-cellulose column chromatography. The dopa decarboxylase formed by the in vitro experiments was characterized by immunoprecipitation and acrylamide gel electrophoresis. It was shown that young larvae that have not yet been exposed to the action of ecdysone contain 3 to 4 times less dopa decarboxylase mRNA than those that have been exposed. This is in good agreement with the levels of dopa decarboxylase itself, the amount of which in the induced larvae is also 3 to 4 times higher than in the uninduced (FRAGOULIS and SEKERIS, 1975b). The results support the concept that ecdysone, like other steroid hormones, induces protein synthesis by increasing the formation of specific mRNAs.

E. Conclusions

It was the intention of the authors to present a general picture of secondary metabolism as part of cell specialization based on the concept of differentiation outlined in the Introduction. Surveying the whole field, the following conclusions may be drawn:

1. A prerequisite for sound studies of secondary metabolism and differentiation is detailed knowledge of the chemistry and biochemistry of secondary products, i.e., of their chemical structures and of the methods for their isolation and quantitation, of the biosynthesis from primary metabolic precursors including the isolation and chemical synthesis of the intermediates required for biosynthetic and enzymatic investigations, and, finally, of the enzymes involved in the biosynthesis, with the aim of reliably determining enzyme activity in vitro.

2. The main regulatory principle determining the expression of secondary metabolism is the de novo synthesis of the enzymes catalyzing the formation of secondary products. Synthesis of these enzymes generally seems to be regulated at the transcriptional level. There are, however, examples in which regulation proceeds also at the translational level (cf. Chap. D2) and by activation of biologically inactive proteinogens (cf. Chap. D1b and D8).

In most experimental systems, the role of differential gene expression was shown by investigating the influence of inhibitors on the dynamics of secondary product formation and on those of the enzymes involved. Considering the manifold methods for studying the regulation of differential gene expression (cf. Chap. A2), the almost exclusive application of inhibitors of gene expression in secondary metabolism is barely satisfactory. Labeling the enzymes with isotopes (cf. Chap. C4), quantifying the corresponding mRNAs, their in vitro translation, and selective determination of the enzyme proteins after immunoprecipitation (cf. Chap. C4 and D8) indicate the direction to be taken in the future.

3. Besides the formation of enzymes, the rate of secondary product synthesis is influenced by mechanisms regulating enzyme activity (e.g., by the levels and compartmentalization of precursors, cosubstrates, end products). These regulatory features, which are not the main theme of this book, are complemented by the influence of regulatory proteins on the amount of biologically active enzymes (cf. Chap. C4) and by mechanisms that repress the de novo formation or increase the degradation of enzymes. Besides a number of general effects, e.g., decline of nucleic acid and protein synthesis or even cell death, specific mechanisms, such as catabolite repression (cf. Chap. D1d) and end product repression (cf. Chap. C4), may decrease secondary product formation.

4. There are direct and indirect results indicating that the enzymes of secondary metabolism form groups with coordinated regulation (cf. Chap. B). However, these groups do not necessarily comprise all enzymes of a particular metabolic chain (cf., e.g., Chap. D1b) and may be dissimilar at different developmental stages (cf. Chap. B).

5. Many low molecular weight compounds are known to effect the expression of secondary metabolism (cf. Chap. C). These include: substrates, products and related compounds (substrate-like effectors: cf. Chap. A1), and hormones as well as other, often chemically unknown, natural factors, which belong to the group

of nonsubstrate-like effectors. None of the effectors investigated so far was shown to influence directly the expression of secondary metabolism. On the contrary, most of them act on the differentiation programs, which include the formation of enzymes of secondary metabolism.

6. One of the most important problems is the integration of secondary metabolism into programs of differentiation and development. This leads to phase-dependent expression of secondary product formation (cf. Chap. D). In many microbial cultures and developing tissues of higher plants and animals a period of rapid cell multiplication (trophophase) is followed by a period of cell specialization (idiophase). Expression of secondary metabolism is only a small part of the metabolic peculiarities expressed during cell specialization in the idiophase. Hence, studies of the phase dependence of secondary metabolism in fact concern the triggering and procedure of often very complex differentiation programs. At present we are far from a detailed understanding of these programs in any one of the experimental systems discussed, though in some instances many facts are already known (cf. the sporulation of *Bacillus* species, Chap. D2, the influence of ecdysone on blowfly pupation, Chap. D8, or the idiophase development of *P. cyclopium*, Chap. D1). It is likely that decisive progress in the elucidation of the programs depends on the isolation and genetic characterization of suitable mutants. However, the genetics of secondary metabolism and the genetics of specialization characteristics in general are, for several reasons, still in their infancy.

Acknowledgements: The authors gratefully acknowledge the permanent experimental efforts of the following students and colleagues who have contributed in a manifold way to the results concerning the chemical, biogenetic, and regulatory aspects of alkaloid metabolism in *P. cyclopium*: S. ABOUTABL, E. BARTSCH, R. DUNKEL, A. EL AZZOUNY, S. EL KOUSY, J. FRAMM, G. ININGER, W. LERBS, W. MÜLLER, H. RICHTER, W. ROOS, I. SCHMIDT, P. SCHRÖDER, S. VOIGT, S. WILSON, K. WINTER. Furthermore, we are grateful to those colleagues who provided us with partly unpublished results and wish to thank Mrs. S. LERBS, Mrs. G. LUCKNER, and Mrs. G. REINBOTHE for their help with the preparation of the manuscript.

References

ABOUTABL, S.A. EL, EL AZZOUNY, A., WINTER, K., LUCKNER, M.: Stereochemical aspects of the conversion of cyclopeptine into dehydrocyclopeptine by cyclopeptine dehydrogenase from Penicillium cyclopium. Phytochemistry 15, 1925-1928 (1976).
ABOUTABL, S.A. EL, LUCKNER, M.: Cyclopeptine dehydrogenase in Penicillium cyclopium. Phytochemistry 14, 2573-2577 (1975).
ACTON, G.J., SCHOPFER, P.: Control over activation or synthesis of phenylalanine ammonia-lyase by phytochrome in mustard (Sinapis alba L.). Biochim. Biophys. Acta 404, 231-242 (1975).
ALFERMANN, W., REINHARD, E.: Isolierung anthocyanhaltiger und anthocyanfreier Gewebestämme von Daucus carota: Einfluß von Auxin auf die Anthocyanbildung. Experientia 27, 353-354 (1971).
ALFERMANN, W., REINHARD, E.: Influence of phytohormones on secondary product formation in plant cell cultures. In: Secondary Metabolism and Coevolution. LUCKNER, M., MOTHES, K., NOVER, L. (eds.). Nova Acta Leopoldina, Suppl. 7, 345-356 (1976).
AMRHEIN, N.: Evidence against the occurrence of adenosine-3',5'-cyclic monophosphate in higher plants. Planta (Berlin) 118, 241-258 (1974).
AMRHEIN, N., ZENK, M.H.: Concomitant induction of phenylalanine ammonia-lyase and cinnamic acid 4-hydroxylase during illumination of excised buckwheat hypocotyls. Naturwissenschaften 57, 312 (1970).
ASHIDA, M., OHNISHI, E.: Activation of pre-phenol oxidase in hemolymph of the silkworm, Bombyx mori. Arch. Biochem. Biophys. 122, 411-416 (1967).
ASHWORTH, J.M.: Cell development in the cellular slime mould Dictyostelium discoideum. Symp. Soc. Exp. Biol. 25, 27-49 (1971).
ATTRIDGE, T.H., FRENCH, C.J., SMITH, H.: The photocontrol of phenylalanine ammonia-lyase levels in Gherkin seedlings - Activation or de novo synthesis? In: Mechanisms of Regulation of Plant Growth. BIELSKI, R.L., FERGUSON, A.F., CRESSWELL, M.M. (eds.). Royal Society of New Zealand: Wellington 1974a, Bull. 12, pp. 361-364.
ATTRIDGE, T.H., JOHNSON, C.B., SMITH, H.: Density-labelling evidence for the phytochrome-mediated activation of phenylalanine ammonia-lyase in mustard cotyledons. Biochim. Biophys. Acta 343, 440-451 (1974b).
ATTRIDGE, T.H., SMITH, H.: Evidence for a pool of inactive phenylalanine ammonia-lyase in Cucumis sativus seedlings. Phytochemistry 12, 1569-1574 (1973).
BAUTZ, E.K.F.: Regulation of RNA synthesis. Progr. Nucl. Acid Res. Molec. Biol. 12, 129-160 (1972).
BECKWITH, J., ROSSOW, P.: Analysis of genetic regulatory mechanisms. Ann. Rev. Genetics 8, 1-13 (1974).
BEERMANN, W. (ed.): Developmental studies on giant chromosomes. Res. Probl. Cell Diff. Berlin-Heidelberg-New York: Springer 1972, Vol. 4.
BELL, E.: Informational DNA synthesis distinguished from that of nuclear DNA by inhibitors of DNA synthesis. Science 174, 603-606 (1971).
BELLINI, E., MARTINELLI, M.: Anthocyanin synthesis in radish seedlings: Effects of continuous far red irradiation and

phytochrome transformations. Z. Pflanzenphysiol. 70, 12-21 (1973).
BERTRAND, K., KORN, L., LEE, F., PLATT, T., SQUIRES, C.L., SQUIRES, C., YANOFSKY, C.: New features of the regulation of the try operon. Science 189, 22-26 (1975).
BETZ, J., TRÄGER, L.: Enzyminduktion bei Streptomyces hydrogenans. Qualitative and quantitative Änderungen des RNA-gehaltes und der RNA-synthese während der Induktion. Z. physiol. Chem. 356, 357-366 (1975).
BÖHM, H.: Über Papaver bracteatum Lindl., 8. Mitt.: Erneutes Auftreten von Alpinigenin in Thebain-Typen nach Verfütterung von Tetrahydropalmatin. Biochem. Physiol. Pflanz. 162, 474-477 (1971).
BOIME, I., BOGUSLAWSKI, S., CAINE, J.: The translation of a human placental lactogen mRNA fraction in heterologous cell-free systems: The synthesis of a possible precursor. Biochem. Biophys. Res. Commun. 62, 103-109 (1975).
BONNER, J.T.: Aggregation and differentiation in the cellular slime moulds. Ann. Rev. Microbiol. 25, 75-92 (1971).
BOULTER, D., ELLIS, R.J., YARWOOD, A.: Biochemistry of protein synthesis in plants. Biol. Rev. 47, 113-175 (1972).
BOYD, G.S., BROWNIE, A.C., JEFCOATE, C.R., SIMPSON, E.R.: Cholesterole hydroxylation in the adrenal cortex and liver. Biochem. J. 125, 1P-2P (1971).
BRAWERMAN, G.: Eukaryotic messenger RNA. Ann. Rev. Biochem. 43, 621-642 (1974).
BREINDL, M., HOLLAND, J.J.: Coupled in vitro transcription and translation of vesicular stomatitis virus messenger RNA. Proc. Nat. Acad. Sci. 72, 2545-2549 (1975).
BRIARTY, L.G., COULT, D.A., BOULTER, D.: Protein bodies of developing seeds of Vicea faba. J. Exp. Botany 20, 358-372 (1969).
BRIGGS, W.R., RICE, H.V.: Phytochrome, chemical and physical properties and mechanism of action. Ann. Rev. Plant Physiol. 23, 293-334 (1972).
BROOKS, J., GRANT, P.R., MUIR, M., GIJZEL, P. VAN, SHAW, G. (eds.):Sporopollenin. London: Academic Press 1971.
BU'LOCK, J.D., BARR, J.G.: A regulation mechanism linking tryptophan uptake and synthesis with ergot alkaloid synthesis in Claviceps. Lloydia 31, 342-354 (1968).
BU'LOCK, J.D., DRAKE, D., WINSTANLEY, D.J.: Specificity and transformations of the trisporic acid series of fungal sex hormones. Phytochemistry 11, 2011-2018 (1972).
BU'LOCK, J.D., HAMILTON, D., HULME, M.A., POWELL, A.J., SMALLEY, H.M., SHEPHERD, D., SMITH, G.N.: Metabolic development and secondary biosynthesis in Penicillium urticae. Can. J. Microbiol. 11, 765-778 (1965).
BU'LOCK, J.D., SHEPHERD, D., WINSTANLEY, D.J.: Regulation of 6-methyl salicylate and patulin synthesis in Penicillium urticae. Can. J. Microbiol. 15, 279-282 (1969).
BURG, S.P.: Ethylene in plant growth. Proc. Nat. Acad. Sci. 70, 591-597 (1973).
BURGEFF, H.: Untersuchungen über Sexualität und Parasitismus bei Mucorineen I. Botan. Abhandl. 4, 1-135 (1924).
BURNETT, A.L.: The acquisition, maintenance, and lability of the differentiated state in Hydra. In: The Stability of the Differentiated State. URSPRUNG, H. (ed.). Berlin-Heidelberg-New York: Springer 1968, Vol. 1, pp. 109-127.

CALHOUN, D.H., HATFIELD, G.W.: Autoregulation of gene expression. Ann. Rev. Microbiol. 29, 275-299 (1975).
CAMM, E.L., TOWERS, G.H.N.: Phenylalanine ammonia-lyase. Phytochemistry 12, 961-973 (1973).
CARSIOTIS, M., JONES, R.F., LACY, A.M., CLEARY, T.J., FRANKHAUSER, D.B.: Histidine-mediated control of tryptophan biosynthetic enzymes in Neurospora crassa. J. Bacteriol. 104, 98-106 (1970).
CASHEL, M.: Regulation of bacterial ppGpp and pppGpp. Ann. Rev. Microbiol. 29, 301-318 (1975).
CLEVER, U.: The effect of ecdysone on gene activity patterns in giant chromosomes. In: Mechanisms of Hormone Action. KARLSON, P. (ed.). Stuttgart: Thieme and New York: Academic Press 1965, pp. 142-148.
CLEVER, U., KARLSON, P.: Induktion von Puff-Veränderungen in den Speicheldrüsenchromosomen von Chironomus tentans durch Ecdyson. Exp. Cell Res. 20, 623-626 (1960).
CLUTTERBUCK, A.J.: The effect of morphological and spore colour mutants on a spore specific enzyme of Aspergillus nidulans. Genetics 24, 515 (1969).
COHEN, M.E., HAMILTON, T.H.: Effect of estradiol-17β on the synthesis of specific uterine nonhistone chromosomal proteins. Proc. Nat. Acad. Sci. 72, 4346-4350 (1975).
COHEN, P.P., BROWN, G.W.: Ammonia metabolism and urea biosynthesis. Comparat. Biochem. 2, 161-214 (1960).
COLOMAS, J., BULARD, C.: Irradiations à faible énergie et biosynthèse d'amarantine chez des plantules d'Amaranthus tricolor L. var. tricolor ruber Hort. Planta (Berlin) 124, 245-254 (1975).
CONSTABEL, F., SHYLUK, J.P., GAMBORG, O.L.: The effect of hormones on anthocyanin accumulation in cell cultures of Haplopappus gracilis. Planta (Berlin) 96, 306-316 (1971).
COOKE, R.J., SAUNDERS, P.F.: Phytochrome-mediated changes in extractable gibberellin activity in a cell-free system from etiolated wheat leaves. Planta (Berlin) 123, 299-302 (1975).
COOKE, R.J., SAUNDERS, P.F., KENDRICK, R.E.: Red light induced production of gibberellin-like substances in homogenates of etiolated wheat leaves and in suspensions of intact etioplasts. Planta (Berlin) 124, 319-328 (1975).
COOTE, J.G., MANDELSTAM, J.: Use of constracted double mutants for determining the temporal order of expression of sporulation genes in Bacillus subtilis. J. Bacteriol. 114, 1254-1263 (1973).
CORRIGAN, J.J.: Nitrogen metabolism in insects. In: Comparative Biochemistry of Nitrogen Metabolism. CAMPBELL, J.W. (ed.). London-New York: Academic Press 1970, Vol. 1, pp. 387-488.
COVE, D.J., PATEMAN, J.A.: Autoregulation of the synthesis of nitrate reductase in Aspergillus nidulans. J. Bacteriol. 97, 1374-1378 (1969).
CRAKER, L.E.: Effect of ethylene and metabolic inhibitors on anthocyanin biosynthesis. Phytochemistry 14, 151-153 (1975).
CRAKER, L.E., ABELES, F.B., SHROPSHIRE, W.: Light-induced ethylene production in Sorghum. Plant Physiol. 51, 1082-1083 (1973).
CRAKER, L.E., WETHERBEE, P.J.: Ethylene, carbon dioxide, and anthocyanin synthesis. Plant Physiol. 52, 177-179 (1973).

CREASY, L.L., ZUCKER, M.: Phenylalanine ammonia-lyase and phenolic metabolism. In: Metabolism and Regulation of Secondary Plant Products. RUNECKLES, V.C., CONN, E.E. (eds.). Rec. Advan. Phytochemistry 8, 1-19 (1974).

CYBIS, J., WEGLEŃSKI, P.: Arginase induction in Aspergillus nidulans. The appearance and decay of the coding capacity of messenger. Europ. J. Biochem. 30, 262-268 (1972).

CZAPEK, F.: Biochemie der Pflanzen. Jena: Fischer 1921, Vol. III, p. 220.

CZICHI, U., KINDL, H.: A model of closely assembled consecutive enzymes on membranes: Formation of hydroxycinnamic acids from L-phenylalanine on thylakoids of Dunaliella marina. Hoppe-Seyler's Z. physiol. Chem. 356, 475-485 (1975).

DARMON, M., BRACHET, P., PEREIRA DA SILVA, L.H.: Chemotactic signals induce cell differentiation in Dictyostelium discoideum. Proc. Nat. Acad. Sci. 72, 3163-3166 (1975).

DAVIS, W.W., GARREN, L.D.: On the mechanism of action of adrenocorticotrophic hormone. The inhibitory site of cycloheximide in the pathway of steroid biosynthesis. J. Biol. Chem. 243, 5153-5157 (1968).

DEMAIN, A.L.: Regulatory mechanisms and the industrial production of microbial metabolites. Lloydia 31, 395-418 (1968).

DEMAIN, A.L.: Overproduction of microbial metabolites and enzymes due to alteration of regulation. Advan. Biochem. Engineering 1, 113-142 (1971).

DEMAIN, A.L.: Cellular and environment factors affecting the synthesis and excretion of metabolites. J. appl. Chem. Biotechnol. 22, 345-362 (1972).

DEMPSEY, M.E.: Regulation of steroid biosynthesis. Ann. Rev. Biochem. 43, 967-990 (1974).

DI CIOCCIO, R.A., STRAUSS, N.: Patterns of transcription in Bacillus subtilis during sporulation. J. Mol. Biol. 77, 325-336 (1973).

DICKSON, R.C., ABELSON, J., BARNES, W.M., REZNIKOFF, W.S.: Genetic regulation: the lac control region. Science 187, 27-35 (1975).

DIENSTMAN, S.R., HOLTZER, H.: Myogenesis: a cell lineage interpretation. In: Cell Cycle and Cell Differentiation. BEERMANN, W., HOLTZER, H., REINERT, J., URSPRUNG, H. (eds.). Berlin-Heidelberg-New York: Springer 1975, pp. 1-25.

DRUMM, H., WILDERMANN, A., MOHR, H.: The "high-irradiance response" in anthocyanin formation as related to the phytochrom level. Photochem. Photobiol. 21, 269-273 (1975).

DRUMMOND, G.J., GREENGARD, P., ROBISON, G.A. (eds.): Advances in Cyclic Nucleotide Research. New York: Raven Press 1975.

DUFFY, J.J., PETRUSEK, R.L., GEIDUSCHEK, E.P.: Conversion of Bacillus subtilis RNA polymerase activity in vitro by a protein induced by phage SPO1. Proc. Nat. Acad. Sci. 72, 2366-2370 (1975).

DUNKEL, R., MÜLLER, W., NOVER, L., LUCKNER, M.: Stimulation of alkaloid formation in Penicillium cyclopium WESTLING by phenylalanine and mycelial extracts. In: Secondary Metabolism and Coevolution. LUCKNER, M., MOTHES, K., NOVER, L. (eds.). Nova Acta Leopoldina, Suppl. 7, 281-288 (1976).

EBEL, J., SCHALLER-HEKELER, B., KNOBLOCH, K.-H., WELLMANN, E., GRISEBACH, H., HAHLBROCK, K.: Coordinated changes in enzyme activities of phenylpropanoid metabolism during the growth

of soybean cell suspension cultures. Biochim. Biophys. Acta 362, 417-424 (1974).
ELGIN, S.C.R., WEINTRAUB, H.: Chromosomal proteins and chromatin structure. Ann. Rev. Biochem. 44, 725-774 (1975).
EL KOUSY, S., PFEIFFER, E., ININGER, G., ROOS, W., NOVER, L., LUCKNER, M.: Influence of inhibitors of gene expression on processes of cell specialization during the idiophase development of Penicillium cyclopium WESTLING. Biochem. Physiol. Pflanz. 168, 79-85 (1975).
ENGELSBERG, E., WILCOX, G.: Regulation: Positive control. Ann. Rev. Genetics 8, 219-242 (1974).
ENGELSMA, G.: Photoinduction of phenylalanine deaminase in gherkin seedlings. III. Effects of excision and irradiation on enzyme development in hypocotyl segments. Planta (Berlin) 82, 355-368 (1968).
ENGELSMA, G.: On the mechanism of the changes in phenylalanine ammonia-lyase activity induced by ultraviolet and blue light in Gherkin hypocotyls. Plant Physiol. 54, 702-705 (1974).
ERGE, D., MAIER, W., GRÖGER, D.: Untersuchungen über die enzymatische Umwandlung von Chanoclavin-I. Biochem. Physiol. Pflanz. 164, 234-247 (1973).
FAILLA, D., TOMKINS, G.M., SANTI, D.V.: Partial purification of a glucocorticoid receptor. Proc. Nat. Acad. Sci. 72, 3849-3852 (1975).
FALK, M., WOLLMANN, R.: Differenzierungsprozesse und Entwicklungsphasen bei Penicillium cyclopium WESTLING - Morphologisch-cytologische Untersuchungen. Pharmazie 29, 77 (1974).
FAN, H., PENMAN, S.: Regulation of protein synthesis in mammalian cells. 2. Inhibition of protein synthesis at the level of initiation during mitosis. J. Mol. Biol. 50, 655-670 (1970).
FEOFILOVA, E.P., BEKHTEREVA, M.N.: Effect of vitamin A on biosynthesis of carotene by Blakeslea trispora. Mikrobiologija 45, 557-558 (1976).
FILNER, P.: Control of nutrient assimilation, a growth-regulating mechanism in cultured plant cells. Develop. Biology, Suppl. 3, 206-226 (1969).
FILNER, P., WRAY, J.L., VARNER, J.E.: Enzyme induction in higher plants. Science 165, 358-367 (1969).
FLOSS, H.G., MOTHES, U.: Über den Einfluß von Tryptophan und analogen Verbindungen auf die Biosynthese von Clavinalkaloiden in saprophytischer Kultur. Arch. Mikrobiol. 48, 213-221 (1964).
FLOSS, H.G., ROBBERS, J.E., HEINSTEIN, P.F.: Regulatory control mechanisms in alkaloid biosynthesis. In: Metabolism and Regulation of Secondary Plant Products. RUNECKLES, V.C., CONN, E.E. (eds.). New York: Academic Press, Recent Advan. Phytochem. 8, 141-175 (1974).
FOARD, D.E.: Differentiation in plant cells. In: Cell Differentiation. SCHJEIDE, O.A., DE VELLIS, J. (eds.). New York: van Nostrand Reinhold Comp. 1970, pp. 575-602.
FOOR, F., JANSSEN, K.A., MAGASANIK, B.: Regulation of synthesis of glutamine synthetase by adenylated glutamine synthetase. Proc. Nat. Acad. Sci. 72, 4844-4848 (1975).
FORGET, B.G., HOUSMAN, D., BENZ, E.J., McCAFFREY, R.P.: Synthesis of DNA complementary to separated human alpha and beta globin messenger RNAs. Proc. Nat. Acad. Sci. 72, 984-988 (1975).
FORRESTER, P.I., GAUCHER, G.M.: m-Hydroxybenzyl alcohol dehydrogenase from Penicillium urticae. Biochemistry 11, 1108-1114 (1972).

FOURNIER, M.J., BRENNER, D.J., DOCTOR, B.P.: The isolation of genes: a review of advances in the enrichment, isolation, and in vitro synthesis of specific cistrons. Progr. Molec. Subcell. Biol. 3, 15-84 (1973).

FRAGOULIS, E.G., SEKERIS, C.E.: Induction of dopa (3,4-dihydroxyphenylalanine) decarboxylase in blowfly integument by ecdysone. A demonstration of synthesis of the enzyme de novo. Biochem. J. 146, 121-126 (1975a).

FRAGOULIS, E.G., SEKERIS, C.E.: Translation of m-RNA for 3,4-dihydroxyphenylalanine decarboxylase isolated from Epidermis tissue of Calliphora vicina R.-D. in an heterologous system. Dependence of m-RNA concentration on the insect steroid hormone ecdysone. Europ. J. Biochem. 51, 305-316 (1975b).

FRAMM, J., NOVER, L., EL AZZOUNY, A., RICHTER, H., WINTER, K., WERNER, S., LUCKNER, M.: Cyclopeptin und Dehydrocyclopeptin, Zwischenprodukte der Biosynthese von Alkaloiden der Cyclopenin-Viridicatin-Gruppe bei Penicillium cyclopium WESTLING. Europ. J. Biochem. 37, 78-85 (1973).

FRENCH, C.J., SMITH, H.: An inactivator of phenylalanine ammonia-lyase from gherkin hypocotyls. Phytochemistry 14, 963-966 (1975).

FREY-WYSSLING, A.: Die Stoffausscheidung der höheren Pflanzen. Berlin: Springer 1935, p. 378.

FREY-WYSSLING, A.: Betrachtungen über pflanzliche Stoffelimination. Ber. Schweiz. Botan. Ges. 80, 454-466 (1970).

FRIEDEN, E., JUST, J.J.: Hormonal responses in amphibian metamorphosis. In: Biochemical Actions of Hormones. LITWACK, G. (ed.). New York: Academic Press 1970, pp. 1-52.

GALSKY, A.G., LIPPINCOTT, J.A.: Promotion and inhibition of α-amylase production in barley endosperm by cyclic AMP and ADP. Plant Cell Physiol. 10, 607-620 (1969).

GILBERT, W., MAXAM, A.: The nucleotide sequence of the lac operator. Proc. Nat. Acad. Sci. 70, 3581-3584 (1973).

GILL, R., VINCE, D.: Photocontrol of anthocyanin formation in turnip seedlings. Planta (Berlin) 86, 116-123 (1969).

GIUDICI DE NICOLA, M., AMICO, V., SCIUTO, S., PIATTELLI, M.: Light control of amaranthin synthesis in isolated Amaranthus cotyledons. Phytochemistry 14, 479-481 (1975).

GIUDICI DE NICOLA, M., PIATTELLI, M., AMICO, V.: Photocontrol of betaxanthin synthesis in Celosia plumosa seedlings. Phytochemistry 12, 353-357 (1973).

GODCHAUX, W., ADAMSON, S.D., HERBERT, E.: Effects of cycloheximide on polyribosome function in reticulocytes. J. Mol. Biol. 27, 57-72 (1967).

GOLDBERGER, R.F.: Autogenous regulation of gene expression. Science 183, 810-816 (1974).

GOODAY, G.W.: Fungal sex hormones. Ann. Rev. Biochem. 43, 35-49 (1974).

GRANICK, S.: The heme and chlorophyll biosynthetic chain. In: Biochemistry of Chloroplasts. GOODWIN, T.W. (ed.). New York: Academic Press 1967, Vol. II, pp. 373-410.

GREENGARD, O.: Enzymic differentiation in mammalian tissues. Essays in Biochemistry 7, 159-205 (1971).

GREENLEAF, A.L., LINN, T.G., LOSICK, R.: Isolation of a new RNA polymerase-binding protein from sporulating Bacillus subtilis. Proc. Nat. Acad. Sci. 70, 490-494 (1973).

GRISEBACH, H.: Neuere Untersuchungen zur Biosynthese einiger Antibiotica. Planta Med. (Stutt.) Supp. 232-250 (1975).
GRISEBACH, H., HAHLBROCK, K.: Enzymology and regulation of flavonoid and lignin biosynthesis in plants and plant cell cultures. In: Metabolism and Regulation of Secondary Plant Products. RUNECKLES, V.C., CONN, E.E. (eds.). New York: Academic Press 1974, pp. 21-52.
GROS, R.: Control of gene expression in prokaryotic systems. FEBS Letters 40, (Suppl.) 19-27 (1974).
GUERZONI, M.E.: Physiological and enzymatic aspects of histidine-mediated control of the tryptophan pathway. Arch. Mikrobiol. 86, 57-68 (1972).
GUIDOTTI, A., HANBAUER, I., COSTRA, E.: Role of cyclic nucleotides in the induction of tyrosine hydroxylase. Advan. Cycl. Nucl. Res. 5, 619-639 (1975).
HAAVIK, H.I., FRØYSHOV, Ø.: Function of peptide antibiotics in producer organism. Nature (London) 254, 79-82 (1975).
HAFFKE, S.C., SEEDS, N.W.: Neuroblastoma: The Escherichia coli of neurobiology? Life Sci. 16, 1649-1658 (1975).
HAHLBROCK, K.: Regulation of phenylalanine ammonia-lyase activity in cell-suspension cultures of Petroselinum hortense. Apparent rates of enzyme synthesis and degradation. Europ. J. Biochem. 63, 137-145 (1976).
HAHLBROCK, K.: Coordinated regulation of the enzymes of flavonoid biosynthesis. In: Cell Differentiation in Microorganisms, Plants and Animals. NOVER, L., MOTHES, K. (eds.). Jena-Amsterdam: VEB Fischer and Elsevier 1977, pp. 524-537.
HAHLBROCK, K., EBEL, J., ORTMANN, R., SUTTER, A., WELLMANN, E., GRISEBACH, H.: Regulation of enzyme activities related to the biosynthesis of flavone glycosides in cell suspension cultures of Parsley (Petroselinum hortense). Biochim. Biophys. Acta 244, 7-15 (1971a).
HAHLBROCK, K., KNOBLOCH, K.-H., KREUZALER, F., POTTS, J.R.M., WELLMANN, E.: Coordinated induction and subsequent activity changes of two groups of metabolically interrelated enzymes. Light-induced synthesis of flavonoid glycosides in cell suspension cultures of Petroselinum hortense. Europ. J. Biochem. 61, 199-206 (1976).
HAHLBROCK, K., RAGG, H.: Light-induced changes of enzyme activities in parsley cell suspension cultures. Effects of inhibitors of RNA and protein synthesis. Arch. Biochem. Biophys. 166, 41-46 (1975).
HAHLBROCK, K., SCHRÖDER, J.: Light-induced changes of enzyme activities in parsley cell suspension cultures. Increased rate of synthesis of phenylalanine ammonia lyase. Arch. Biochem. Biophys. 166, 47-53 (1975).
HAHLBROCK, K., SUTTER, A., WELLMANN, E., ORTMANN, R., GRISEBACH, H.: Relationship between organ development and activity of enzymes involved in flavone glycoside biosynthesis in young parsley plants. Phytochemistry 10, 109-116 (1971b).
HAHLBROCK, K., WELLMANN, E.: Light-induced flavone biosynthesis and activity of phenylalanine ammonia-lyase and UDP-apiose synthetase in cell suspension cultures of Petroselinum hortense. Planta (Berlin) 94, 236-239 (1970).
HAHLBROCK, K., WELLMANN, E.: Light-independent induction of enzymes related to phenylpropanoid metabolism in cell suspension cultures from parsley. Biochim. Biophys. Acta 304, 702-706 (1973).

HAMILTON, T.H.: Control by estrogen of genetic transcription and translation. Science 161, 649-661 (1968).
HAMKALO, B.A., MILLER, O.L.: Electronmicroscopy of genetic activity. Ann. Rev. Biochem. 42, 379-396 (1973).
HANSON, R.S., PETERSON, J.A., YOUSTEN, A.A.: Unique biochemical events in bacterial sporulation. Ann. Rev. Microbiol. 24, 53-90 (1970).
HARTLEY, M.R., WHEELER, A., ELLIS, R.J.: Protein synthesis in chloroplasts. V. Translation of mRNA for the large subunit of fraction I protein in a heterologous cell-free system. J. Mol. Biol. 91, 67-77 (1975).
HATHOUT BASSIM, T.A., PECKET, R.C.: The effect of membrane stabilizers on phytochrome-controlled anthocyanin biosynthesis in Brassica oleracea. Phytochemistry 14, 731-733 (1975).
HAYMAN, E.P., CHICHESTER, C.O., SIMPSON, K.L.: Effects of CPTA upon carotenogenesis and lipoidal constituents in Rhodotorula species. Phytochemistry 13, 1123-1127 (1974).
HEINSTEIN, P.F., LEE, S.-L., FLOSS, H.G.: Isolation of dimethylallylpyrophosphate: Tryptophan dimethylallyl transferase from the ergot fungus (*Claviceps spec.*). Biochem. Biophys. Res. Comm. 44, 1244-1251 (1971).
HESLOP-HARRISON, J.: Sporopollenin in the biological context. In: Sporopollenin. BROOKS, J., GRANT, P.R., MUIR, M., VAN GIJZEL, SHAW, G. (eds.). London: Academic Press 1971, pp. 1-30.
HESS, D.: Die Wirkung von Zimtsäuren auf die Anthocyansynthese in isolierten Petalen von Petunia hybrida. Z. Pflanzenphysiol. 56, 12-19 (1967a).
HESS, D.: Substratinduktion bei der Anthocyansynthese von Petunia. Naturwissenschaften 54, 289-290 (1967b).
HESS, D.: Substratinduktionen durch Zimtsäuren als physiologischer Mechanismus bei der Einleitung der Anthocyansynthese. Z. Pflanzenphysiol. 59, 293-296 (1968).
HESS, D., HADWIGER, L.A.: The induction of phenylalanine ammonialyase and phaseollin by 9-aminoacridine and other deoxyribonucleic acid intercalating compounds. Plant Physiol. 48, 197-202 (1971).
HIPPEL, P.H. VON, McGHEE, J.D.: DNA-protein interactions. Ann. Rev. Biochem. 41, 231-300 (1972).
HOLZER, H.: Regulatory role of proteinases and proteinase inhibitors in yeast. In: Mechanisms of Action and Regulation of Enzymes. KELETI, T. (ed.). Proc. 9th FEBS Meeting 32, 181-193. Budapest: Akadémikai Kiadó 1975.
HOLTZER, H., ABOTT, J.: Oscillation of the chondrogenic phenotype in vitro. In: The Stability of the Differentiated State. URSPRUNG, H. (ed.). Berlin-Heidelberg-New York: Springer 1968, pp. 1-16.
HOLTZER, H., WEINTRAUB, H., BIEHL, J.: Cell cycle-dependent events during myogenesis, neurogenesis, and erythrogenesis. In: Biochemistry of Cell Differentiation. MONROY, A., TSANEV, R. (eds.). London: Academic Press 1973, pp. 41-53.
HOROWITZ, J., SAUKKONEN, J.J., CHARGAFF, E.: Effects of fluoropyrimidines on the synthesis of bacterial proteins and nucleic acids. J. Biol. Chem. 235, 3266-3272 (1960).
HU, A.S.L., BOCK, R.M., HALVORSON, H.O.: Separation of labeled from unlabeled proteins by equilibrium density gradient sedimentation. Anal. Biochem. 4, 489-504 (1962).

HUA, S., MARKOVITZ, A.: Multiple regulation of the galactose operon - genetic evidence for a distinct site in the galactose operon that responds to capR gene regulation in Escherichia coli K-12. Proc. Nat. Acad. Sci. 71, 507-511 (1974).
HUANG, P.C.: DNA, RNA and protein interactions. Progr. Biophys. Mol. Biol. 23, 103-144 (1972).
HUANG, P.C., SMITH, M.M.: Nucleic acid hybridization and the nature of chromosomal protein bound RNA. In: Nucleic Acid Hybridization in the Study of Cell Differentiation. URSPRUNG, H. (ed.). Berlin-Heidelberg-New York: Springer 1972, pp. 65-76.
HUEZ, G., MARBAIX, H., HUBERT, E., LECLERCQ, M., NUDEL, U., SOREQ, H., SALOMON, R., LEBLEU, B., REVEL, M., LITTAUER, U.Z.: Role of the polyadenylate segment in the translation of globin messenger RNA in Xenopus oocytes. Proc. Nat. Acad. Sci. 71, 3143-3146 (1974).
IMAMOTO, F.: Translation and transcription of the tryptophan operon. Progr. Nucl. Acid. Res. Mol. Biol. 13, 339-407 (1973).
INGLE, J.: The effect of light and inhibitors on chloroplast and cytoplasmatic RNA synthesis. Plant Physiol. 43, 1850-1854 (1968).
ININGER, G., NOVER, L.: Regulation of β-galactosidase formation in emerged cultures of Penicillium cyclopium WESTLING. Biochem. Physiol. Pflanz. 167, 585-595 (1975).
IREDALE, S.E., SMITH, H.: Properties of phenylalanine ammonia-lyase extracted from Cucumis sativus hypocotyls. Phytochemistry 13, 575-583 (1974).
JANGAARD, N.O.: The characterization of phenylalanine ammonia-lyase from several plant species. Phytochemistry 13, 1765-1768 (1974).
JANISTYN, B., DRUMM, H.: Phytochrome-mediated rapid changes of cyclic AMP in mustard seedling (Sinapis alba L.). Planta (Berlin) 125, 81-85 (1975).
JENSEN, E.V., DE SOMBRE, E.R.: Estrogen receptor interaction. Science 182, 126-134 (1973).
JENSEN, E.V., NUMATA, M., SMITH, S., SUZUKI, T., BRECHER, P.I., DE SOMBRE, E.R.: Estrogen-receptor interactions in target tissues. Dev. Biol. Suppl. 3, 151-171 (1969).
JICÍNSKÁ, E.: Note on study of the sporulation of fungi: endotrophic sporulation in the genus Penicillium. Folia Microbiol. (Praha) 13, 401-409 (1968).
JOBE, A., BOURGEOIS, S.: Lac repressor-operator interaction; VI: The natural inducer of the lac operon. J. Mol. Biol. 69, 397-408 (1972).
JOH, T.H., GEGHMAN, C., REIS, D.: Immunochemical demonstration of increased accumulation of tyrosine hydroxylase protein in sympathetic ganglia and adrenal medulla elicited by reserpine. Proc. Nat. Acad. Sci. 70, 2767-2771 (1973).
JOHNSON, C., ATTRIDGE, T., SMITH, H.: Regulation of phenylalanine ammonia-lyase synthesis by cinnamic acid; its implication of the light mediated regulation of the enzyme. Biochim. Biophys. Acta 385, 11-19 (1975).
JONDORF, W.R., SIMON, D.C., AVNIMELECH, M.: Further studies on the stimulation of L-(^{14}C)-amino acid incorporation with cycloheximide. Mol. Pharmacol. 2, 506-517 (1966).
JONES, G.H., WEISSBACH, H.: RNA metabolism in Streptomyces antibioticus; Effect of 5-fluorouracil on the appearance of phenoxazinone synthetase. Arch. Biochem. Biophys. 137, 558-573 (1970).

JOST, J.P., RICKENBERG, H.V.: Cyclic AMP. Ann. Rev. Biochem. 40, 741-774 (1971).
JUNGMANN, R.A., SHAW-GUANG LEE, DeANGELO, A.B.: Translocation of cytoplasmic protein kinase and cAMP-binding protein to intracellular acceptor sites. Advan. Cycl. Nucl. Res. 5, 281-306 (1975).
KAMBE, M., IMAE, Y., KURABASHI, K.: Biochemical studies on gramicidin S non-producing mutants of Bacillus brevis ATCC 9999. J. Biochem. (Tokyo) 75, 481-493 (1974).
KANE, J.F., HOLMES, W.M., JENSEN, R.A.: Metabolic interlock: The dual function of a folate pathway gene as an extra-operonic gene of tryptophan biosynthesis. J. Biol. Chem. 247, 1587-1596 (1972).
KANG, B.G., NEWCOMB, W., BURG, S.P.: Mechanism of auxin-induced ethylene production. Plant Physiol. 47, 504-509 (1971).
KANNANGARA, C.G., HENNINGSEN, K.W., STUMPF, P.K., WETTSTEIN, D. VON: 6-Methylsalicylic acid synthesis by isolated barley chloroplasts. Europ. J. Biochem. 21, 334-338 (1971).
KAPLAN, H., HORNEMANN, U., KELLEY, K.M., FLOSS, H.G.: Tryptophan metabolism, protein and alkaloid synthesis in saprophytic cultures of the ergot fungus (Claviceps sp.). Lloydia 32, 489-497 (1969).
KARABOYAS, G.C., KORITZ, S.B.: Identity of the site of action of 3',5'-adenosine monophosphate and adrenocorticotrophic hormone in corticosteroidogenesis in rat adrenal and beef adrenal cortex slices. Biochemistry 4, 462-468 (1965).
KARLSON, P.: Hormonwirkung durch Genaktivierung. In: Mechanisms of Hormone Action. KARLSON, P. (ed.). Stuttgart: Thieme and New York: Academic Press 1965, pp. 139-141.
KARLSON, P., MERGENHAGEN, D., SEKERIS, C.E.: Zum Tyrosinstoffwechsel der Insekten, XV. Weitere Untersuchungen über das o-Diphenoloxydasesystem von Calliphora erythrocephala. Z. Physiol. Chem. 338, 42-50 (1964).
KARLSON, P., SCHWEIGER, A.: Zum Tyrosinstoffwechsel der Insekten, IV. Das Phenoloxydase-System von Calliphora und seine Beeinflussung durch das Hormon Ecdyson. Hoppe-Seyler's Z. Physiol. Chem. 323, 199-210 (1961).
KARLSON, P., SEKERIS, C.E.: Zum Tyrosinstoffwechsel der Insekten, IX. Kontrolle des Tyrosinstoffwechsels durch Ecdyson. Biochim. Biophys. Acta 63, 489-495 (1962).
KARLSON, P., SEKERIS, C.E., MARMARAS, V.I.: Die Aminosäurezusammensetzung verschiedener Proteinfraktionen aus der Cuticula von Calliphora erythrocephala in verschiedenen Entwicklungsstadien. J. Insect. Physiol. 15, 319-323 (1969).
KAU, K.W., UNGAR, F.: Characterization of an adrenal activator for cholesterol side chain cleavage. J. Biol. Chem. 248, 2868-2875 (1973).
KEPES, A.: Transcription and translation in the lactose operon of Escherichia coli studied by in vivo kinetics. Progr. Biophys. Mol. Biol. 19, 199-236 (1969).
KHOKHLOV, A.S., ANISOVA, L.N., TOVAROVA, I.I., KLEINER, E.M., KOVALENKO, I.V., KRASILNIKOVA, O.I., KORNITSKAYA, I.Y., PLINER, S.A.: Effect of A-factor on the growth of asporogenous mutants of Streptomyces griseus, not producing this factor. Z. Allg. Mikrobiol. 13, 647-655 (1973).
KHOKHLOV, A.S., TOVAROVA, I.I., ANISOVA, L.N.: Regulators of streptomycin biosynthesis and development of Actinomyces strep-

tomycini. In: Secondary Metabolism and Coevolution. LUCKNER, M., MOTHES, K., NOVER, L. (eds.). Nova Acta Leopoldina, Suppl. 7, 289-298 (1976).

KHOKHLOV, A.S., TOVAROVA, I.I., BORISOVA, L.N., PLINER, S.A., SHEVCHENKO, L.A., KORNITSKAYA, E.J., IVKINA, N.S., RAPOPORT, I.A.: The A-factor responsible for the biosynthesis of streptomycin in mutant strains of Actinomyces streptomycini. Dokl. Akad. Nauk SSSR 177, 232-235 (1967).

KILLEWICH, L., SCHUTZ, G., FEIGELSON, P.: Functional level of rat liver tryptophan 2,3-dioxygenase messenger RNA during superinduction of enzyme with actinomycin D. Proc. Nat. Acad. Sci. 72, 4285-4287 (1975).

KILLICK, K.A., WRIGHT, B.A.: Regulation of enzyme activity during differentiation in Dictyostelium dioscoideum. Ann. Rev. Microbiol. 28, 139-166 (1974).

KIM, K.-H.: Nucleic acid hybridization to isolated chromatin. In: Nucleic Acid Hybridization in the Study of Cell Differentiation. URSPRUNG, H. (ed.). Berlin-Heidelberg-New York: Springer 1972, pp. 37-46.

KLIER, A.F., LECADET, M.M., DEDONDER, R.: Sequential modifications of DNA-dependent RNA polymerase during sporogenesis in Bacillus thuringiensis. Europ. J. Biochem. 36, 317-327 (1973).

KÖHLER, K.-H.: Kinetik der durch Inhibitoren des Protein- und Nukleinsäurestoffwechsels bewirkten Hemmung der Amaranthinbildung bei Amaranthus caudatus Keimlingen. Biochem. Physiol. Pflanz. 168, 113-122 (1975).

KRUPINSKI, V.M., ROBBERS, J.E., FLOSS, H.G.: Physiological study of ergot: induction of alkaloid synthesis by tryptophan at the enzymatic level. J. Bact. 125, 158-165 (1976).

KUNG, H.-F., SPEARS, C., WEISSBACH, H.: Purification and properties of a soluble factor required for the DNA-directed in vitro synthesis of β-galactosidase. J. Biol. Chem. 250, 1556-1562 (1975).

KURAHASHI, K.: Biosynthesis of small peptides. Ann. Rev. Biochem. 43, 445-459 (1974).

LANGE, H., MOHR, H.: Die Hemmung der Phytochrom-induzierten Anthocyansynthese durch Actinomycin D und Puromycin. Planta (Berlin) 67, 107-121 (1965).

LANGE, H., SHROPSHIRE, W., MOHR, H.: An analysis of phytochrom-mediated anthocyanin synthesis. Plant Physiol. 47, 649-655 (1971).

LANE, M.D., MOSS, J.: Regulation of fatty acid synthesis in animal tissues. In: Metabolic Regulation. VOGEL, H.J. (ed.). New York-London: Academic Press 1971, pp. 23-54.

LAWRENCE, P.: The cell cycle and cellular differentiation in insects. In: Cell Cycle and Cell Differentiation. BEERMANN, W., HOLTZER, H., REINERT, J., URSPRUNG, H. (eds.). Berlin-Heidelberg-New York: Springer 1975, pp. 111-121.

LEE, S.G., LITTAU, V., LIPMANN, F.: The relation between sporulation and the induction of antibiotic synthesis and of amino acid uptake in Bacillus brevis. J. Cell Biol. 66, 233-242 (1975).

LEIGHTON, T.J., DOI, R.H., WARREN, R.A.I., KELLN, R.A.: The relationship of serine protease activity to RNA polymerase modification and sporulation in Bacillus subtilis. J. Mol. Biol. 76, 103-122 (1973).

LIEBERMANN, M., KUNISHI, A.T.: Ethylene-forming systems in etiolated pea seedling and apple tissue. Plant Physiol. 55, 1074-1078 (1975).
LIGHT, R.J.: Effects of cycloheximide and amino acid analogues on biosynthesis of 6-methylsalicylic acid in Penicillium patulum. Arch. Biochem. Biophys. 122, 494-500 (1967).
LIGHT, R.J.: 6-Methylsalicylic acid decarboxylase from Penicillium patulum. Biochim. Biophys. Acta 191, 430-438 (1969).
LIGHT, R.J.: Enzymatic studies on polyketide hypothesis. J. Agr. Food Chem. 18, 260-267 (1970).
LINN, T.G., GREENLEAF, A.L., SHORENSTEIN, R.G., LOSICK, R.: Loss of the σ-activity of RNA polymerase of Bacillus subtilis during sporulation. Proc. Nat. Acad. Sci. 70, 1865-1869 (1973).
LIS, J.T., SCHLEIF, R.: The regulatory region of the L-arabinose operon: its isolation on a 1000 base-pair fragment from DNA heteroduplexes. J. Mol. Biol. 95, 409-416 (1975).
LISSITZKY, S., FAYET, G., VERRIER, B.: Thyrotropin-receptor interaction and cAMP-mediated effects in thyroid cells. Advan. Cycl. Nucl. Res. 5, 133-152 (1975).
LOCKWOOD, D.H., STOCKDALE, F.E., TOPPER, Y.Y.: Hormone dependent differentiation of mammary gland: Sequence of action of hormones in relation to cell cycle. Science 156, 945-946 (1967).
LÖFFELHARDT, W., KINDL, H.: The conversion of L-phenylalanine into benzoic acid on the thylakoid membrane of higher plants. Hoppe-Seyler's Z. Physiol. Chem. 356, 487-493 (1975).
LÖFFELHARDT, W., LUDWIG, B., KINDL, H.: Thylakoid-gebundene L-Phenylalanin-Ammoniak-Lyase. Hoppe Seyler's Z. Physiol. Chem. 354, 1006-1012 (1973).
LOSICK, R.: In vitro transcription. Ann. Rev. Microbiol. 41, 409-446 (1972).
LOSICK, R., PERO, J.: Bacillus subtilis RNA polymerase and its modification in sporulating and phage-infected bacteria. Adv. Enzymol. 44, 165-185 (1976).
LUCKNER, M.: Was ist Sekundärstoffwechsel? Pharmazie 26, 717-724 (1971).
LUCKNER, M.: Secondary Metabolism in Plants and Animals. London: Chapman and Hall Ltd. 1972.
LUCKNER, M.: The integration of benzodiazepine and quinoline alkaloid formation into the developmental program of Penicillium cyclopium. In: Cell Differentiation in Microorganisms, Plants and Animals. NOVER, L., MOTHES, K. (eds.). Jena and Amsterdam: VEB Fischer and Elsevier 1977, pp. 538-558.
LUCKNER, M., MOTHES, K., NOVER, L.: Secondary metabolism and coevolution; cellular, intercellular and interorganismic aspects. Nova Acta Leopoldina, Suppl. 7 (1976).
MACKENZIE, J.M., COLEMAN, R.A., BRIGGS, W.R., PRATT, L.H.: Reversible redistribution of phytochrome within the cell upon conversion to its physiologically active forms. Proc. Nat. Acad. Sci. 72, 799-803 (1975).
MAGASANIK, B.: Regulation of the synthesis of degradative bacterial enzymes. In: Cell Differentiation in Microorganisms, Plants and Animals. NOVER, L., MOTHES, K. (eds.). Jena and Amsterdam: VEB Fischer and Elsevier 1977, pp. 438-460.
MAGASANIK, B., PRIVAL, M.J., BRENCHLEY, J.E., TYLER, B.M., DE LEO, A.B., STREICHER, S.L., BENDER, R.A., PARIS, C.G.: Glutamine synthetase as a regulator of enzyme synthesis. Current Topics Cell. Regul. 8, 119 (1974).

MAHAFFEE, E., REITZ, R.C., NEY, R.L.: The mechanism of action of adrenocorticotrophic hormone. J. Biol. Chem. 249, 227-233 (1974).
MAILHAMMER, R., YANG, H.L., REINESS, G., ZUBAY, G.: Effects of bacteriophage T4-induced modification of Escherichia coli RNA polymerase on gene expression in vitro. Proc. Nat. Acad. Sci. 72, 4928-4932 (1975).
MAJORS, J.: Initiation of in vitro mRNA synthesis from the wild-type lac promoter. Proc. Nat. Acad. Sci. 72, 4394-4398 (1975).
MANCINELLI, A.L., YANG, C.H., LINDQUIST, P., ANDERSON, O.R., RABINO, I.: Photocontrol of anthocyanin synthesis. III. The action of streptomycin on the synthesis of chlorophyll and anthocyanin. Plant Physiol. 55, 251-257 (1972).
MANDELSTAM, J.: Bacterial sporulation: a problem in the biochemistry and genetics of a primitive developmental system. Proc. Roy Soc. London B. 193, 89-106 (1976).
MARMARAS, V.J., SEKERIS, C.E.: Pattern and activity of polyribosomes from the integument of blowfly larvae during development. Exp. Cell Res. 75, 143-153 (1972).
MARSHALL, R., REDFIELD, B., KATZ, E., WEISSBACH, H.: Changes in phenoxazinone synthetase activity during the growth cycle of Streptomyces antibioticus. Arch. Biochem. Biophys. 123, 317-323 (1968).
MARTINELLI, S.D.: Biochemical investigations on conidiation of Aspergillus nidulans in submerged liquid culture. Biochem. J. 127, 16 P (1972).
MATCHETT, W.H., DEMOSS, J.A.: Physiological channeling of tryptophan in Neurospora crassa. Biochim. Biophys. Acta 86, 91-99 (1964).
McCLURE, J.W.: Physiology and function of flavonoids. In: The Flavonoids. HARBORNE, J.B., MABRY, T.J., MABRY, H. (eds.). London: Chapman and Hall 1975, pp. 970-1055.
McCLURE, J.W., GROSS, G.G.: Diverse photoinduction characteristics of hydroxycinnamate: Coenzyme A ligase and phenylalanine ammonia-lyase in dicotyledonous seedlings. Z. Pflanzenphysiol. 76, 51-55 (1975).
McMAHON, D.: Cycloheximide is not a specific inhibitor of protein synthesis in vivo. Plant Physiol. 55, 815-821 (1975).
MEINS, F.: Cell division and the determination phase of cytodifferentiation in plants. In: Cell Cycle and Cell Differentiation. BEERMANN, W., HOLTZER, H., REINERT, J., URSPRUNG, H. (eds.). Berlin-Heidelberg-New York: Springer 1975, pp. 151-175.
MESLAND, D.A.M., HUISMAN, J.G., VAN DEN ENDE, H.: Volatile sexual hormones in Mucor mucedo. J. Gen. Microbiol. 80, 111-117 (1974).
METZENBERG, R.L.: Genetic regulatory systems in Neurospora. Ann. Rev. Genetics 6, 111-132 (1972).
MEYER, B.J., KLEID, D.G., PTASHNE, M.: λ Repressor turns off transcription of its own gene. Proc. Nat. Acad. Sci. 72, 4785-4789 (1975).
MITRAKOS, K., SHROPSHIRE, W. (eds.): Phytochrome. London: Academic Press 1972.
MOHR, H., DRUMM, H., KASEMIR, H.: Licht und Farbstoffe. Ber. Deut. Botan. Ges. 87, 49-69 (1974).
MOHR, H., SITTE, P.: Lectures on Photomorphogenesis. Berlin-Heidelberg-New York: Springer 1972.
MOSCONA, A.A.: Induction of glutamine synthetase in embryonic neural retina: a model for the regulation of specific gene

expression in embryonic cells. In: Biochemistry of Cell Differentiation. MONROY, A., TSANEV, R. (eds.). London-New York: Academic Press 1973, pp. 1-23.

MOSCONA, M., FRENKEL, N., MOSCONA, A.A.: Regulatory mechanisms in the induction of glutamine synthetase in the embryonic retina: Immunochemical studies. Develop. Biol. 28, 229-241 (1972).

MOTHES, K.: Vergleichende Betrachtung des pflanzlichen Stoffwechsels. In: Physiologische Chemie. FLASCHENTRÄGER, B., LEHNARTZ, E. (eds.). Berlin-Heidelberg-New York: Springer 1966a, Vol. 11/2d/B, pp. 971-975.

MOTHES, K.: Zur Problematik der metabolischen Exkretion bei Pflanzen. Naturwissenschaften 53, 317-323 (1966b).

MOTHES, K.: Über sekundäre Pflanzenstoffe. Abh. sächs. Akad. Wiss. Leipzig, math.-naturwiss. Klasse 52, 3-29 (1972).

MÜNTZ, K., HORSTMANN, C., SCHOLZ, G.: Proteine und Proteinbiosynthese in Samen von Vicia faba L. Kulturpflanze 20, 277-326 (1972).

MURPHY, G., LYNEN, F.: Patulin biosynthesis. The metabolism of m-hydroxybenzyl alcohol and m-hydroxybenzaldehyde by particulate preparations from Penicillium patulum. Europ. J. Biochem. 58, 467-475 (1975).

NASON, A., EVANS, H.J.: TPN-nitrate reductase in Neurospora: Inducibility of nitrate reductase. J. Biol. Chem. 202, 655-673 (1953).

NEBERT, D.W., GIELEN, J.E., GOUJON, F.M.: Genetic expression of aryl hydrocarbon hydroxylase induction. 3. Changes in the binding of n-octylamine to cytochrome P 450. Mol. Pharmacol. 8, 651-666 (1972).

NEGBI, M., HOPKINS, D.W., BRIGGS, W.R.: Acceleration of dark reversion of phytochrome in vitro by calcium and magnesium. Plant Physiol. 56, 157-159 (1975).

NOVER, L., MÜLLER, W.: Influence of cycloheximide on the expression of alkaloid metabolism in partially synchronized emerged cultures of Penicillium cyclopium WESTLING. FEBS Letters 50, 17-20 (1975).

NOVER, L., LUCKNER, M.: Mixed functional oxygenations during the biosynthesis of cyclopenin and cyclopenol, benzodiazepine alkaloids of Penicillium cyclopium WESTLING. FEBS Letters 3, 292-296 (1969).

NOVER, L., LUCKNER, M.: Expression of secondary metabolism as part of the differentiation processes during the idiophase development of Penicillium cyclopium WESTLING. Biochem. Physiol. Pflanz. 166, 293-305 (1974).

NOVER, L., LUCKNER, M.: Influence of inhibitors of gene expression on processes of cell specialization in Penicillium cyclopium WESTLING. In: Secondary Metabolism and Coevolution. LUCKNER, M., MOTHES, K., NOVER, L. (eds.). Nova Acta Leopoldina, Suppl. 7, 229-241 (1976).

OELZE-KAROW, H., MOHR, H.: Experiments regarding the problem of differentiation in multicellular systems. Z. Naturforsch. 25b, 1282-1286 (1970).

O'MALLEY, B.W., MEANS, A.R.: Female steroid hormones and target cell nuclei. Science 183, 610-620 (1974).

ORNSTON, L.N.: Regulation of catabolic pathways in Pseudomonas. Bact. Rev. 35, 87-116 (1971).

OTA, Y., OHTA, T., IMAHORI, K.: The differentiation in the availabilities of initiation points of phage RNA cistrons in their translation. J. Biochem. 71, 743-746 (1972).
PACKTER, N.M., COLLINS, J.S.: Effect of inhibitors of protein synthesis on the formation of phenols derived from acetate and shikimic acid in Aspergillus fumigatus. Europ. J. Biochem. 42, 291-302 (1974).
PAECH, K.: Biochemie und Physiologie der sekundären Pflanzenstoffe. Berlin-Göttingen-Heidelberg: Springer 1950.
PALADE, G.E.: Intracellular aspects of the process of protein synthesis. Science 189, 347-358 (1975).
PALMITER, R.D., CAREY, N.H.: Rapid inactivation of ovalbumin messenger ribonucleic acid after acute withdrawel of estrogen. Proc. Nat. Acad. Sci. 71, 2357-2361 (1974).
PARTHIER, B: Existenz und Realisierung extrachromosomaler genetischer Information in Plastiden und Mitochondrien. Biol. Rundschau 8, 289-306 (1970).
PARTHIER, B., KRAUSPE, R., MUNSCHE, D., WOLLGIEHN, R.: The biogenesis of chloroplasts. In: The Chemistry and Biochemistry of Plant Proteins. HARBORNE, J.B., VAN SUMERE, C.F. (eds.). New York-London: Academic Press 1975, pp. 167-270.
PASTAN, I., PERLMAN, R.: Cyclic adenosine monophosphate in bacteria. Science 169, 339-344 (1970).
PAUL, J., GILMOUR, R.S.: The regulatory role of non-histone proteins in RNA synthesis. Ciba Foundation Symp. 28, 181-198 (1975).
PAULUS, H., SARKAR, N.: Peptide antibotics as regulatory effectors during bacterial sporulation. Nova Acta Leopoldina, Suppl. 7, 357-373 (1976).
PECKET, R.C., HATHOUT BASSIM, T.A.: The effect of kinetin in relation to photocontrol of anthocyanin biosynthesis in Brassica oleracea. Phytochemistry 13, 1395-1399 (1974).
PERO, J., NELSON, J., FOX, T.D.: Highly asymmetric transcription by RNA polymerase containing phage-SPO1-induced polypeptides and a new host protein. Proc. Nat. Acad. Sci. 72, 1589-1593 (1975).
PESTKA, S.: Inhibitors of ribosome functions. Ann. Rev. Biochem. 40, 697-710 (1971); Ann. Rev. Microbiol. 25, 487-562 (1971).
PIATTELLI, M., GIUDICI DE NICOLA, M., CASTROGIOVANNI, V.: The Effect of kinetin on amaranthin synthesis in Amaranthus tricolor in darkness. Phytochemistry 10, 289-293 (1971).
PIGGOT, P.J.: Mapping of asporogenous mutations of Bacillus subtilis: a minimum estimate of the number of sporulation operons. J. Bacteriol. 114, 1241-1253 (1973).
PINE, M.J.: Turnover of intracellular proteins. Ann. Rev. Microbiol. 26, 103-126 (1972).
PIRROTTA, V.: Sequence of the O_R operator phage λ. Nature (London) 254, 114-117 (1975).
POLING, S.M., HSU, W.-J., YOKOYAMA, H.: Structure-activity relationships of chemical inducers of carotenoid biosynthesis. Phytochemistry 14, 1933-1938 (1975).
POYTON, R.O., GROOT, G.S.P.: Biosynthesis of polypeptides of cytochrome C oxidase by isolated mitochondria. Proc. Nat. Acad. Sci. 72, 172-176 (1975).
QUAST, L., WIERMANN, R.: Über das Vorkommen verschieden substituierter Chalkone während der Mikrosporogenese bei Tulipa. Experientia 29, 1165-1166 (1973).

RABINOWITCH, H.D., RUDICH, J.: Effects of ethephon and CPTA on color development of tomato fruits at high temperature. Hort. Sci. 7, 76-77 (1972).
RANJEVA, R., BOUDET, A.M., HARADA, H., MARIGO, G.: Phenolic metabolism in petunia tissues. I. Characteristic responses of enzymes involved in different steps of polyphenol synthesis to different hormonal influences. Biochim. Biophys. Acta 399, 23-30 (1975).
RAO, K.K., PATEL., V.P.: Effect of tryptophan and related compounds on alkaloid formation in Aspergillus fumigatus. Lloydia 37, 608-610 (1974).
RAST, D., SKŘIVANOVÁ, R., BACHOFEN, R.: Replacement of light by dibutyryl-cAMP and cAMP in betacyanin synthesis. Phytochemistry 12, 2669-2672 (1973).
REXER, B., SRINIVASAN, V.R., ZILLIG, W.: Regulation of transcription during sporulation of Bacillus cereus T. Europ. J. Biochem. 53, 271-281 (1975).
RHAESE, H.-J., GROSCURTH, R.: Control of development: Role of regulatory nucleotides synthesized by membranes of Bacillus subtilis in initiation of sporulation. Proc. Nat. Acad. Sci. 73, 331-335 (1976).
RICHMOND, M.H.: Enzymic adaptation in bacteria: its biochemical and genetic basis. Essays in Biochemistry 4, 105-154 (1968).
RICHTER, I., LUCKNER, M.: Cyclopenin m-hydroxylase - an enzyme of alkaloid metabolism in Penicillium cylcopium. Phytochemistry 15, 67-70 (1976).
RICKENBERG, H.V.: Cyclic AMP in prokaryotes. Ann. Rev. Microbiol. 28, 353-369 (1974).
RICKWOOD, D., BIRNIE, G.A.: Metrizamide, a new density-gradient medium. FEBS Letters 50, 102-110 (1975).
RISTOW, H.-J., SCHAZSCHNEIDER, B., BAUER, K., KLEINKAUF, H.: Tyrocidine and the linear gramicidin. Do these peptide antibiotics play an antagonistic regulative role in sporulation? Biochim. Biophys. Acta 390, 246-252 (1975a).
RISTOW, H.-J., SCHAZSCHNEIDER, B., BAUER, K.: Effects of the peptide antibiotics tyrocidine and the linear gramicidin on RNA synthesis and sporulation of Bacillus brevis. Biochem. Biophys. Res. Commun. 63, 1085-1092 (1975b).
RISTOW, H.-J., SCHAZSCHNEIDER, B., VATER, J., KLEINKAUF, H.: Some characteristics of the DNA-tyrocidine complex and a possible mechanism of the gramicidin action. Biochim. Biophys. Acta 414, 1-8 (1975c).
ROBBERS, J.E., FLOSS, H.G.: Induction by tryptophan of the enzymes of ergot alkaloid biosynthesis. Nova Acta Leopoldina, Suppl. 7, 243-269 (1976).
ROBBERS, J.E., ROBERTSON, L.W., HORNEMANN, K.M., JINDRA, A., FLOSS, H.G.: Physiological studies on ergot: Further studies on the induction of alkaloid synthesis by tryptophan and its inhibition by phosphate. J. Bacteriol. 112, 791-796 (1972).
SAIER, M.H., FEUCHT, B.U., McCAMAN, M.T.: Regulation of intracellular adenosine cyclic 3',5'-monophosphate levels in Escherichia coli and Salmonella typhimurium. J. Biol. Chem. 250, 7593-7601 (1975).
SAKAI, S., IMASEKI, H.: Auxin-induced ethylene production by mungbean hypocotyl segments. Plant Cell Physiol. 12, 349-359 (1971).

SARKAR, N., PAULUS, H.: Function of peptide antibiotics in sporulation. Nature (New Biol.) 239, 228-230 (1972).
SARKAR, P.K., MOSCONA, A.A.: Glutamine synthetase induction in embryonic neural retina: Immunochemical identification of polysomes involved in enzyme synthesis. Proc. Nat. Acad. Sci. 70, 1667-1671 (1973).
SAUNDERS, J.A., McCLURE, J.W.: Phytochrome controlled phenylalanine ammonia lyase in Hordeum vulgare plastids. Phytochemistry 14, 1285-1289 (1975).
SCHÄFER, E.: Evidence for binding of phytochrome to membranes. In: Membrane Transport in Plants. ZIMMERMANN, U., DAINTY, J. (eds.). Berlin-Heidelberg-New York: Springer 1974, pp. 435-440.
SCHAZSCHNEIDER, B., RISTOW, H., KLEINKAUF, H.: Interaction between the antibiotic tyrocidine and DNA in vitro. Nature (London) 249, 757-759 (1974).
SCHERRER, K.: Formation and regulation of mRNA in animal cells. In: Regulation of transcription and Translation in Eukaryotes. BAUTZ, E.K.F., KARLSON, P., KERSTEN, H. (eds.). Berlin-Heidelberg-New York: Springer 1973, pp. 81-104.
SCHIFF, J.A.: The control of chloroplast differentiation in Euglena. In: Proc. 3rd Int. Congr. Photosynthesis. AVRON, M. (ed.). Amsterdam: Elsevier 1974, pp. 1691-1717.
SCHIMKE, R.T.: Control of enzyme levels in mammalian tissues. Advan. Enzymol. 37, 135-187 (1973).
SCHOPFER, P., BAJRACHARYA, D., FALK, H., THIEN, W.: Phytochromgesteuerte Entwicklung von Zellorganellen (Plastiden, Microbodies, Mitochondrien). Ber. Deut. Botan. Ges. 88, 245-268 (1975).
SCHOPFER, P., PLACHY, C.: Determination by phytochrome of enzyme development. Nova Acta Leopoldina, Suppl. 7, 327-334 (1976).
SCHULSTER, D.: Adrenocorticotrophic hormone and the control of adrenal corticosteroidogenesis. Advan. Steroid Biochem. Pharmacol. 4, 233-295 (1974).
SCHWEIGER, A., KARLSON, P.: Zum Tyrosinstoffwechsel der Insekten, X. Die Aktivierung der Präphenoloxydase und das Aktivator-Enzym. Z. Physiol. Chem. 329, 210-221 (1962).
SEITZ, U., HEINZMANN, U.: Einfluß der Gibberellinsäure A_3 auf die Anthocyansynthese der Kalluskulturen von Daucus carota. Planta Med. (Stuttgart) Suppl. 66-69 (1975).
SEKERIS, C.E.: Action of ecdysone on RNA and protein metabolism in the blowfly, Calliphora erythrocephala. In: Mechanisms of Hormone Action. KARLSON, P. (ed.). Stuttgart: Thieme and New York: Academic Press 1965, pp. 149-167.
SEKERIS, C.E.: Die Wirkung von Hormonen auf den Zellkern. Chemie in unserer Zeit 3, 171-177 (1969).
SEKERIS, C.E., KARLSON, P.: On the mechanism of hormone action, II. Ecdysone and protein biosynthesis. Arch. Biochem. Biophys. 105, 483-487 (1964).
SEKERIS, C.E., MERGENHAGEN, D.: Phenoloxidase system of the blowfly, Calliphora erythrocephala. Science 145, 68-69 (1964).
SERCARZ, E.E., GORINI, L.: Different contribution of exogenous and endogenous arginine to repressor formation. J. Mol. Biol. 8, 254-262 (1964).
SEYAMA, N., SPLITTSTOESSER, W.E.: Pigment synthesis in Cucurbita moschata cotyledons as influenced by CPTA and several inhibitors. Plant Cell Physiol. 16, 13-19 (1975).

SHAAYA, E., SEKERIS, C.E.: Inhibitory effects of α-amanitin on RNA synthesis and induction of DOPA-decarboxylase by β-ecdysone. FEBS Letters 16, 333-336 (1971).
SHAPIRO, D.J., SCHIMKE, R.T.: Immunochemical isolation and characterization of ovalbumin messenger ribonucleic acid. J. Biol. Chem. 250, 1759-1764 (1975).
SHAW, G.: The chemistry of sporopollenin. In: Sporopollenin. BROOKS, J., GRANT, P.R., MUIR, M., VAN GIJZEL, P., SHAW, G. (eds.). London: Academic Press 1971, pp. 305-348.
SHOWE, M.K., DE MOSS, J.A.: Localization and regulation of nitrate reductase in Escherichia coli. J. Bacteriol. 95, 1305-1313 (1968).
SILVER, M.J., SMITH, J.B.: Prostaglandins as intracellular messengers. Life Sci. 16, 1635-1648 (1975).
SMILLIE, R.M., SCOTT, N.S.: Organelle biosynthesis: The chloroplast. Progr. Mol. Subcell. Biol. 1, 136-202 (1969).
SMITH, H.: The photocontrol of flavonoid biosynthesis. In: Phytochrome. MITRAKOS, K., SHROPSHIRE, W. (eds.). London: Academic Press 1972, pp. 433-481.
SMITH, H.: Regulatory mechanisms in the photocontrol of flavonoid biosynthesis. In: Biosynthesis and Its Control in Plants. MILBORROW, B.V. (ed.). London-New York: Academic Press 1973, pp. 303-321.
SONDHEIMER, E., SIMEONE, J.B. (eds.): Chemical Ecology. New York: Academic Press 1970.
STAFFORD, H.A.: Possible multienzyme complexes regulating the formation of C_6-C_3 phenolic compounds and lignins in higher plants. In: Metabolism and Regulation of Secondary Plant Products. RUNECKLES, V.C., CONN, E.E. (eds.). New York: Academic Press 1974, pp. 53-79.
STANLEY, R.G., LINSKENS, H.F.: Pollen: Biology, Biochemistry, Management. Berlin-Heidelberg-New York: Springer 1974.
STEFFENSEN, D.M., WIMBER, D.E.: Hybridization of nucleic acids to chromosomes. In: Nucleic Acid Hybridization in the Study of Cell Differentiation. URSPRUNG, H. (ed.). Berlin-Heidelberg-New York: Springer 1972, pp. 47-63.
STEPHENS, J.C., ARZT, S.W., AMES, B.N.: Guanosine 5'-diphosphate-3'-diphosphate (ppGpp): positive effector for histidine operon transcription and general signal for amino-acid deficiency. Proc. Nat. Acad. Sci. 72, 4389-4393 (1975).
STERLINI, J.M., MANDELSTAM, J.: Commitment to sporulation in Bacillus subtilis and its relationship to development of actinomycin resistance. Biochem. J. 113, 29-37 (1969).
SUGIYAMA, T., KORANT, B.D., LONBERG-HOLM, K.K.: RNA virus gene expression and its control. Ann. Rev. Microbiol. 26, 468-502 (1972).
SÜTFELD, R., WIERMANN, R.: Die Bildung von Coenzym A-Thiolestern verschieden substituierter Zimtsäuren durch Enzymextrakte aus Antheren. Z. Pflanzenphysiol. 72, 163-171 (1974).
SUTHERLAND, E.W.: Studies on the mechanism of hormone action. Science 177, 401-408 (1972).
SUTTER, R.P.: Mutations affecting sexual development in Phycomyces blakesleeanus. Proc. Nat. Acad. Sci. 72, 127-130 (1975).
SZYBALSKI, W.: A network of developmental controls in coliphage lambda. In: Cell Differentiation in Microorganisms, Plants and Animals. NOVER, L., MOTHES, K. (eds.). Jena and Amsterdam: VEB Fischer and Elsevier 1977, pp. 262-316.

SZYBALSKI, W., BOVRE, K., FIANDT, M., HAYES, S., HRADECNA, Z., KUMAR, S., LOZERON, H.H., NIJKAMP, H.J.J., STEVENS, W.F.: Transcriptional units and their control in Escherichia coli phage λ: Operons and scriptons. Cold Spring Harbor Symp. Quant. Biol. 35, 341-354 (1970).
TATA, J.R.: Regulation of protein synthesis by growth and developmental hormones. In: Biochemical Actions of Hormones. LITWACK, G. (ed.). New York: Academic Press 1970, pp. 89-133.
THIEN, W., SCHOPFER, P.: Control by phytochrome of cytoplasmic and plastid rRNA accumulation in cotyledons of mustard seedlings in the absence of photosynthesis. Plant Physiol. 56, 660-664 (1975).
THOMAS, D.M., HARRIS, R.C., KIRK, J.T.O., GOODWIN, T.W.: Studies on carotenogenesis in Blakeslea trispora, II. The mode of action of trisporic acid. Phytochemistry 6, 361-366 (1967).
THOMAS, R.L., JEN, J.J.: Phytochrome-mediated carotenoid biosynthesis in ripening tomatoes. Plant Physiol. 50, 452-453 (1975).
TOMKINS, G.M., GELEHRTER, T.D., GRANNER, D., MARTIN, D., SAMUELS, H.H., THOMPSON, E.B.: Control of specific gene expression in higher organisms. Science 166, 1474-1480 (1969).
TOMKINS, G.M., LEVINSON, B.B., BAXTER, J.D., DETZLEFSEN, L.: Further evidence for posttranscriptional control of inducible tyrosine aminotransferase synthesis in cultures of hepatoma cells. Nature (New Biol.) 239, 9-14 (1972).
TOVAROVA, I.I., KORNITSKAYA, E.J., PLINER, C.A., SHEVCHENKO, L.A., ANISOVA, L.N., KHOKHLOV, A.S.: On the role of the A-factor in the biosynthesis of streptomycin. Izvest. Akad. Nauk SSSR, Ser. Biol. No. 3, 427-434 (1970).
TRAN THANH VAN, M., CHLYAH, H., CHLYAH, A.: Regulation of organogenesis in thin layers of epidermal and sub-epidermal cells. In: Tissue Culture and Plant Science. STREET, H.E. (ed.). London: Academic Press 1974, pp. 101-139.
TSAI, M.-J., O'MALLEY, B.W.: Effects of estrogen on gene expression in chick oviduct. In: Cell Differentiation in Microorganisms, Plants and Animals. NOVER, L., MOTHES, K. (eds.). Jena and Amsterdam: VEB Fischer and Elsevier 1977, pp. 109-125.
TUAN-HUA HO, D., VARNER, J.E.: Hormonal control of messenger ribonucleic acid metabolism in barley aleurone layers. Proc. Nat. Acad. Sci. 71, 4783-4786 (1974).
TYLER, B., DELEO, A., MAGASANIK, B.: Activation of transcription of hut DNA by glutamine synthetase. Proc. Nat. Acad. Sci. 71, 225-229 (1974).
UMBARGER, H.E.: Regulation of amino acids metabolism. Ann. Rev. Biochem. 38, 323-370 (1969).
VAN DEN ENDE, H., WERKMAN, B.A., VAN DEN BRIEL, M.L.: Trisporic acid synthesis in mated cultures of the fungus Blakeslea trispora. Arch. Mikrobiol. 86, 175-184 (1972).
VAN DEN ENDE, H., WIECHMAN, A.H.C.A., REYNGOUD, D.J., HENDRIKS, T.: Hormonal interactions in Mucor mucedo and Blakeslea trispora. J. Bacteriol. 101, 423-428 (1970).
VAUGHAN, M.H., PAWLOWSKI, P.J., FORCHNAMMER, J.: Regulation of protein synthesis initiation in HeLa cells deprived of single essential amino acids. Proc. Nat. Acad. Sci. 68, 2057-2061 (1971).
VERMA, D.P.S., MACLACHLAN, G.A., BYRNE, H., EWINGS, D.: Regulation and in vitro translation of m-RNA for cellulase from auxin-treated pea epicotyls. J. Biol. Chem. 250, 1019-1026 (1975).

VOGEL, H.J. (ed.): Metabolic Regulation. New York-London: Academic Press 1971.
VOIGT, S., LUCKNER, M.: Dehydrocyclopeptine epoxidase, a mixed function oxygenase of the alkaloid metabolism of Penicillium cyclopium. Phytochemistry, in press.
WAITES, W.M., KAY, D., DAWES, I.W., WOOD, D.A., WARREN, S.C., MANDELSTAM, J.: Sporulation in Bacillus subtilis. Correlation of biochemical events with morphological changes in asporogenous mutants. Biochem. J. 118, 667-676 (1970).
WALTON, D.C., SOOFI, G.S., SONDHEIMER, E.: The effects of abscissic acid on growth and nucleic acid synthesis in excised embryonic bean axes. Plant Physiol. 45, 37-40 (1970).
WALZ, A., PIRROTTA, V.: Sequence of the P_R promoter of phage λ. Nature (London) 254, 118-121 (1975).
WARD, A.C., PACKTER, N.M.: Relationship between fatty-acid and phenol synthesis in Aspergillus fumigatus. Europ. J. Biochem. 46, 323-333 (1974).
WEINBERG, E.D.: Biosynthesis of secondary metabolites: Role of trace metals. Advan. Microbiol. Physiol. 4, 1-44 (1970).
WEINBERG, E.D., TONNIS, S.M.: Action of chloramphenicol and its isomers on secondary biosynthetic processes of Bacillus. Appl. Microb. 14, 850-856 (1966).
WEINBERG, E.D., TONNIS, S.M.: Role of manganese in biosynthesis of bacitracin. Can. J. Microb. 13, 614-615 (1967).
WEINTRAUB, H.: The organization of red cell differentiation. In: Cell Cycle and Cell Differentiation. BEERMANN, W., HOLTZER, H., REINERT, J., URSPRUNG, H. (eds.). Berlin-Heidelberg-New York: Springer 1975, pp. 27-42.
WEISENSEEL, M., HAUPT, W.: The photomorphogenic pigment phytochrome: a membrane effector? In: Membrane Transport in Plants. ZIMMERMANN, U., DAINTY, J. (eds.). Berlin-Heidelberg-New York: Springer 1974, pp. 427-434.
WELLMANN, E.: Regulation der Flavonoidbiosynthese durch ultraviolettes Licht und Phytochrom in Zellkulturen und Keimlingen von Petersilie (Petroselinum hortense Hoffm.). Ber. Deut. Botan. Ges. 87, 267-273 (1974).
WELLMANN, E.: UV dose-dependent induction of enzymes related to flavonoid biosynthesis in cell suspension cultures of parsley. FEBS Letters 51, 105-107 (1975).
WELLMANN, E., SCHOPFER, P.: Phytochrome mediated de novo synthesis of phenylalanine ammonia-lyase in cell suspension cultures of parsley. Plant Physiol. 55, 822-827 (1975).
WERKMAN, T.A., VAN DEN ENDE, H.: Trisporic acid synthesis in Blakeslea trispora. Interaction between plus and minus mating types. Arch. Mikrobiol. 90, 365-374 (1973).
WIERMANN, R.: Über die Beziehungen zwischen flavonolaufbauenden Enzymen, einem flavonolumwandelnden Enzym und der Akkumulation phenylpropanoider Verbindungen während der Antherenentwicklung. Planta (Berlin) 110, 353-360 (1973).
WILCOX, G., MEURIS, P., BASS, R., ENGLESBERG, E.: Regulation of the L-arabinose operon BAD in vitro. J. Biol. Chem. 249, 2946-2952 (1974).
WILKINSON, D.S., PITOT, H.C.: Inhibition of ribosomal RNA maturation in Novikoff hepatoma cells by 5-fluorouracil and 5-fluorouridine. J. Biol. Chem. 248, 63-68 (1973).
WILSON, S., SCHMIDT, I., ROOS, W., FÜRST, W., LUCKNER, M.: Quantitative Bestimmung des Enzyms Cyclopenase in Konidiosporen

von Penicillium cyclopium WESTLING und Penicillium viridicatum WESTLING. Z. Allg. Mikrobiol. 14, 515-523 (1974).
WILSON, S., LUCKNER, M.: Cyclopenase, ein Lipoproteid der Protoplasmamembran von Konidiosporen des Pilzes Penicillium cyclopium WESTLING. Z. Allg. Mikrobiol. 15, 45-51 (1975).
WOO, S.L.C., O'MALLEY, B.W.: Hormone inducible messenger RNA review. Life Sci. 17, 1039-1048 (1975).
WOOD, D.A.: Sporulation in Bacillus subtilis. The appearance of sulpholactic acid as a marker event for sporulation. Biochem. J. 123, 601-605 (1971).
YALOW, R.S., BERSON, S.A.: Immunoassay of plasma insulin. Methods Biochem. Anal. 12, 69-96 (1964).
ZENK, M.H., EL-SHAGI, H., SCHULTE, U.: Anthraquinone production by cell suspension cultures of Morinda citrifolia. Planta Med. (Stuttgart) Suppl. 79-101 (1975).
ZUCKER, M.: Light and enzymes. Ann. Rev. Plant Physiol. 23, 133-156 (1972).

II. Secondary Metabolism in Cell Cultures of Higher Plants and Problems of Differentiation
HARTMUT BÖHM

A. Introduction

Progress in understanding differentiation in higher organisms depends on two conditions: (1) the knowledge that differentiation is a process at the cellular level, and (2) studies of metabolism carried out in suitable systems.

There is no reason to discuss the first condition because of the constructive concept outlined by LUCKNER and NOVER in the foregoing article. However, the author does not wish to neglect the reactions initiated by proteins during gene expression. It appears reasonable from a biologic point of view that a cell is characterized by all products of differentiation processes irrespective of their properties. This opinion is expressed also in the following chapters.

Differentiation processes do not always lead to morphologic features - many of them are detectable only at the metabolic level. At present this so-called biochemical differentiation allows more successful experiments than the formation of specific cellular structures. It makes possible elucidation of several differentiation steps, which probably have general significance.

Secondary metabolism in plants offers an excellent opportunity for studying differentiation problems (cf. LUCKNER and NOVER). The production of secondary substances comprises biosynthesis, transport, accumulation, and storage. Frequently, these processes occur in individual cells (for example flavonoid and tannin metabolism), but in some cases biosynthesis and storage are distributed in different cells. Differentiation of cells participating in secondary metabolism may lead to high specialization, which is indicated by loss of normal cell functions.

The search for the most suitable systems for the investigation of secondary metabolism resulted in a preference for cultivated plant cells rather than intact plants. As a rule, cell cultures allow both the simple detection of metabolic reactions and extensive modification of experimental conditions. On the other hand, it is well documented that the formation of secondary metabolites observed in the intact plant is often lacking in the derived cell culture. This situation, however, may be favorable for experiments directed toward problems of differentation because of the possibility of inducing specific processes.

The purpose of this acticle is to treat the subject by offering a selection of pertinent experimental results and a discussion of unsolved questions. The general term "cell culture" is used exclusively throughout the article, but for correct interpretation of experimental data the original terms are symbolized by capitals (C = callus culture, T = tissue culture, S = cell suspension culture). For more information about the fundamental questions of biochemical differentiation and plant cell culture, the reader is referred to two excellent publications: the paper of KRIKORIAN and STEWARD (1969) and the book edited by STREET (1973), respectively.

B. The Fate of Secondary Metabolism during Initiation of Plant Cell Cultures

Secondary metabolism is indeed the result of differentiation processes, but the presence of secondary substances is not necessarily indicative of the simultaneous involvement of gene expressions in this situation. Apart from the possibility that the formation of alkaloids, pigments, volatile oils, and others is catalyzed by stable enzymes for some time without de novo synthesis of proteins, the storage of such substances is a special problem. Whether or not the storage of secondary metabolites requires specific genetic information is unknown. Thus, a cell may store a pigment, for example, although related gene expressions are absent: those that led to synthesis and accumulation are completed, those that are responsible for storage do not exist. Such a cell, however, differs from cells without pigment; it is differentiated in the sense explained previously (cf. Chap. A). This statment emphasizes that the term "differentiated state" does not represent a biologic unity but rather various states of remarkably different significance for other cellular processes. Under these restrictions the following experiments must be considered.

Beside tissue parts, single cells (e.g. BHATT and MEHTA, 1974) and protoplasts (e.g. PELCHER et al., 1974) from intact plants are also used for initiation of cell cultures. From this fact the first question arises, whether any cell specialized in secondary metabolism continues its function after complete isolation. Protoplasts isolated from petals of flower buds of *Nemesia strumosa* showed anthocyanin synthesis in a 16-hour experiment (HESS and ENDRESS, 1973). Obviously, corresponding studies with cells have not yet been described in the literature. Primary metabolic reactions were proved in isolated cells from tobacco (JENSEN et al., 1971). Considering these findings and the function of isolated protoplasts, one may suppose that differentiated cells continue secondary metabolism for some time after isolation.

Division of single cells and of protoplasts, respectively, is a prerequisite for callus formation from this material. Thus a further question concerns the ability of cells participating in secondary metabolism to divide. Pertinent successful experiments were performed with cells from cell cultures only. Using the plating method CONSTABEL et al. (1971) observed no division of single anthocyanin-containing cells of *Haplopappus gracilis* (S). However, pigmented cells within groups of four or more cells showed cell wall formation and proliferation under otherwise identical conditions. No decrease of the anthocyanin content was detectable during cell division. Alkaloid-accumulating cells from cell cultures (S) of *Macleaya microcarpa*, characterized by yellow color, do not divide even within cell groups. Most of the specialized cells lose their color while other cells of a plated cell population divide. This points to a despecialization before the division of alkaloid-accumulating cells in the *Macleaya* system (KOBLITZ and SCHUMANN, unpubl. data). Nevertheless, in principle, cells specialized in secondary metabolism are able to divide. Loss of the differentiated state is evidently not a prerequisite for this process. The fact that single cells do not or only seldom divide represents a general problem of cell cultivation.

Continuation of secondary metabolism in isolated plant tissues for some hours is well known from biochemical studies. Cultivation of tissue parts on nutrient agar may lead to a remarkable extension of the metabolically active phase. Explants of lemon pericarp show oil passages and a slightly changing content of volatile oil until the onset of proliferation (DE BILLY and PAUPARDIN, 1971). Also the secondary phloem of explanted *Juniperus communis* still contains tannin cells at the time of callus formation. These specialized cells undergo subsequent change, which is especially expressed by dilution of the content of vacuoles. However, no total loss of tannins takes place. Such partly despecialized cells divide. Neither mother nor daughter cells showed tannin synthesis. Tannin synthesizing cells occur only in the peripheral layers of the callus (CONSTABEL, 1968).

Although the cited results indicate favorable conditions for continuous secondary metabolism during initiation of cell cultures, i.e., especially for the start of these metabolic processes in the originating calli, obviously a definite difficulty exists: Frequently the differentiated state in explanted cells does not last until the onset of proliferation. As exemplified by tannin cells in explants from *J. communis* (CONSTABEL, 1968), aside from accumulated secondary substances, the fundamental biosynthetic activity may be absent. Also decreased amounts of secondary metabolites were observed repeatedly in anthocyanin-containing explants (e.g. IBRAHIM et al., 1971). Colorlessness at the end of this process indicates a total loss of specialization. On the other hand, callus of lemon explants showed no oil passages despite the presence of the probably complete secondary metabolism in the starting material at the time of proliferation (DE BILLY and PAUPARDIN, 1971).

Evidently, constant synthesis and accumulation of secondary substances in plant cell cultures arise from differentiation processes that are independent of explants under the given culture conditions. Even if specialized cells turn from explant into callus, they should cause detectable secondary metabolism for a short time only. Certainly, the division rates of specialized cells are lower than those of unspecialized ones, and elimination of the former is the consequence. There are examples showing that secondary metabolism was observed only during the first subcultivations of cells (see IBRAHIM et al., 1971).

C. Realization of Secondary Metabolism in Plant Cell Cultures

1. Triggering Factors

In this field favored by the properties of cell cultures, the triggering of flavonoid metabolism is the area that has been most intensively studied thus far. From intact plants it was known that differentiation processes leading to the synthesis and accumulation of flavonoids generally are light-dependent (GRISEBACH, 1965; GRISEBACH and BARZ, 1969). Triggering of flavone (HAHLBROCK and WELLMANN, 1970; BRUNET and IBRAHIM, 1973) and anthocyanin formation by light (GREGOR and REINERT, 1972; ALFERMANN, 1973) has been shown in cell cultures from several plant species. In these experiments no or only traces of secondary substances were detectable in cells before the onset of light. Illumination did not influence the growth rate of cell cultures. Generally speaking the latter two findings represent criteria of the action of a factor triggering secondary metabolism.

If dark grown cell cultures (S) from *Petroselinum hortense*, for example, are illuminated with white light, 24 glycosides of several flavones and flavonoles occur (HAHLBROCK and WELLMANN, 1970; KREUZALER and HAHLBROCK, 1973). The formation of flavone glycosides begins after a lag phase of four to six hours. Three to four days after the onset of light the concentration of pigments reaches a constant level (HAHLBROCK and WELLMANN, 1970). The developmental stage of cell cultures at the time of illumination is of significance for the course of pigment accumulation: After the onset of light, cell cultures (S) from *H. gracilis* grown in the dark for three days show earlier a start and maximum of anthocyanin accumulation than cells subcultivated for one day in the dark (Fig. 47). In addition, the values of the respective maxima are quite different (FRITSCH et al., 1971). In these experiments blue light was used because of the finding that synthesis of anthocyanin in *H. gracilis* cell cultures (T) depends strictly on light of this quality. Red light of short and long wavelengths is ineffective (REINERT et al., 1964). The action spectrum of anthocyanin synthesis in cell cultures (T) of *H. gracilis* has two peaks, one at 438 nm, the other at 372 nm (LACKMANN, 1971). In *P. hortense* cell cultures (S) wavelengths below

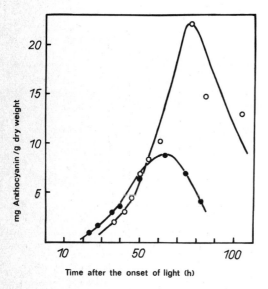

Fig. 47. Course of anthocyanin accumulation in cell cultures (S) from *Haplopappus gracilis*. Before onset of light, cultures were grown in dark for 1 day (-o-) and 3 days (-•-), respectively. (From FRITSCH et al., 1971, redrawn)

320 nm represent the most effective spectral range of the white light used for triggering flavone glycoside synthesis. Therefore, ultraviolet light (UV) alone can induce pigment formation, which shows a linear dependence on UV doses (WELLMANN, 1971, 1975). Red light also is unable to trigger flavonoid metabolism in this system. However, after preirradiation with UV, red light exerts a significant influence on flavone glycoside synthesis in cell cultures (S) of parsley. This result appears remarkable in view of changes in the state of cells occurring after irradiation with UV. Obviously, phytochrome changes from an inactive to an active state (WELLMANN, 1971; WELLMANN and BARON, 1974).

Triggering of differentiation processes by auxins was demonstrated with a line of *Daucus carota* cell cultures (T; ALFERMANN and REINHARD, 1971): Colorless cells obtained by cultivation in the absence of light and auxin showed anthocyanin synthesis about six days after transfer to a medium containing 2,4-dichlorophenoxyacetic acid (2,4-D), indole-3-acetic acid (IAA), or α-naphthaleneacetic acid (NAA) under otherwise identical conditions. In general, light triggers anthocyanin synthesis in *D. carota* cell cultures. Evidently, replacement of light by auxin is possible. However, the course of anthocyanin accumulation after triggering by light and by auxin, respectively, is quite different (ALFERMANN, 1973). It is not yet known whether light and auxin act according to the same principle.

Anthraquinone synthesis in cell cultures (S) from *Morinda citrifolia* depends strictly on auxins. However, of 146 substances showing auxin activity tested, only a few were able to trigger anthraquinone metabolism. NAA was most effective. In contrast, *M. citrifolia* cells cultivated in a medium containing 2,4-D as the only auxin did not produce anthraquinones (ZENK et al., 1975). The dependence of some areas of secondary metabolism on specific auxins was also reported for other systems: Presence of IAA is a prerequisite for naphthoquinone synthesis in *Lithospermum erythrorhizon* cell cultures (C). Under the influence of 2,4-D alone no pigment is formed (TABATA et al., 1974). In cell cultures (C) from *Nicotiana tabacum*, phytosterols and triterpenes but no alkaloids occur on a medium with 2,4-D. If under otherwise identical conditions auxin is represented by IAA, the cells produce alkaloids, but phytosterols and triterpenes are not detectable (FURUYA et al., 1967, 1971). A strain of *Plumbago zeylanica* cell cultures (C) forms anthocyanins on a medium containing 2,4-D. These pigments are not synthesized in the presence of IAA and NAA, respectively, as the only auxin (HEBLE et al., 1974).

Knowledge of triggering factors provides a basis for further experiments on the mode of action of these effectors. In this respect, the cited results point to the surprising exchangeability of factors that obviously are quite different, while the replacement of a triggering substance by a very similar compound may be ineffective. Perhaps different triggering factors lead to different spectra of a given class of secondary products. For example, the anthocyanins detectable in *D. carota* cell cultures (T) under the influence of light and auxin, respectively, could be of different compositions (see SCHMITZ and SEITZ, 1972). There is no doubt that the elucidation of the action of triggering factors is promoted by detection of the resulting gene expression at an earlier step than the synthesis of related secondary substances.

2. Enzyme Activities

Experiments with plant cell cultures have yielded significant information about enzymatic reactions of secondary metabolism. However, only some of the results obtained in this field have importance for the problem of differentiation. In addition, apart from a few cases, the de novo synthesis of a particular enzyme can be concluded only from an increase of enzyme activity.

The enzymes involved in the synthesis of flavone glycosides in cell cultures (S) from *P. hortense* belong to two different groups based on behavior. Group 1 includes L-phenylalanine ammonia-lyase (PAL), cinnamic acid 4-hydroxylase, and p-coumarate:CoA ligase. Generally, these enzymes catalyze the formation of phenylpropane derivatives. In contrast, the enzymes of group 2 are engaged exclusively in flavone glycoside synthesis: flavanone synthetase, glucosyltransferase, apiosyltransferase, UDP-apiose synthetase, and others (cf. Figs. 1 and 2 of LUCKNER and NOVER, p. 11, 12). The activities of all enzymes of groups 1 and 2 show an increase about two and four hours, respectively, after the beginning of illumination of parsley cells previously grown in the dark for ten days. The first enzymes (group 1) reach their maximum activities after 17 to 23 hours, the others (group 2), after 26 to 37 hours only (Fig. 48). Thus, a common regulation of the enzymes of each group and a difference between the two regulatory systems are assumed. As demonstrated with PAL, maximum enzyme activity depends on the developmental stage of the parsley cells at the time of illumination (HAHLBROCK et al., 1971a; SUTTER and GRISEBACH, 1973; HAHLBROCK et al., 1976).

Feeding experiments with methinonine-$[^{35}S]$ (HAHLBROCK and SCHRÖDER, 1975a) and ^{15}N-labeled compounds (WELLMANN and SCHOPFER, 1975) demonstrated the de novo synthesis of PAL in dark grown cell cultures (S) of *P. hortense* after illumination. When inhibitors of transcription and translation were added to parsley cell cultures (S) before the onset of light, the increase of PAL activity was inhibited up to 100% compared with untreated cultures. Under these conditions the activities of p-coumarate:CoA ligase and chalcone-flavanone isomerase were also much lower than in control experiments. Addition of inhibitors to parsley cell cultures (S) at different times after the onset of light influenced the PAL activity in the following way: The inhibitory effect of actinomycin D decreased linearly from the start of illumination, whereas that of cycloheximide remained constant throughout the lag phase and diminished afterwards only. These results demonstrate the sequence of RNA and protein formation during de novo synthesis of PAL (HAHLBROCK and RAGG, 1975).

It is remarkable that dark grown cell cultures (S) of *P. hortense* show activities of all the enzymes involved in flavone glycoside

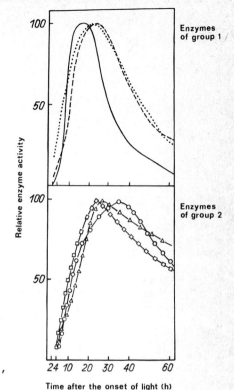

Fig. 48. Activities of various enzymes involved in flavone glycoside biosynthesis in cell cultures (S) from *Petroselinum hortense*: PAL (−), cinnamic acid 4-hydroxylase (−−−), p-coumarate:CoA ligase (...), flavanone synthetase (-○-) glucosyltransferase (-□-), UDP-apiose synthetase (-△-). (From HAHLBROCK et al., 1976, redrawn)

synthesis. The activities of most enzymes amount to less than 10% of those detectable under the influence of light. Only cinnamic acid 4-hydroxylase and p-coumarate:CoA ligase reach higher levels of activities (30-40%) in the dark (HAHLBROCK et al., 1971a). Nevertheless, these findings indicate that the formation of enzymes participating in pigment synthesis in parsley cells does not depend strictly on light. On the other hand, the relatively large amounts of some enzymes and related products are unable to induce adequate pigment synthesis without the influence of the triggering factor. This conclusion is well supported by the following experiments in the dark: When *P. hortense* cells subcultivated for ten days were transferred to fresh liquid medium, the activities of all enzymes belonging to group 1 sharply increased. Starting two hours after the transfer, the increase led to maximum activities after 15 hours with the exception of cinnamic acid 4-hydroxylase activity (HAHLBROCK and WELLMANN, 1973). Similar results were obtained after transfer of cell cultures (T) of several *Citrus* species (THORPE et al., 1971) and of *H. gracilis* (GREGOR and REINERT, 1972) to fresh media in the dark. The increase of PAL activity began some hours after transfer and reached maxima after one day (*H. gracilis*) and two days (*Citrus* species), respectively. The subsequent decrease led during four and two weeks, respectively, to an enzyme activity that corresponded to the original level. Illumination had no or little effect on these processes. Only after some days was blue light able to stimulate PAL activity in transferred cell cultures (T) of *H. gracilis* (GREGOR and REINERT, 1972). It is possible that the unknown "factor" triggering the increase of PAL activity in the dark acts on the same principle as the blue light and prevents its effect. On the other hand, even after transfer of dark grown parsley cells (S) into distilled water a large increase of various enzyme activities occurred. These processes were also restricted to enzymes of group 1. Obviously, they are not induced by a triggering factor but by dilution of at least one compound arising from the cells. As exemplified with PAL, the extent of activity change depends strictly on the degree of dilution of cells in distilled water. Five hours after transfer of parsley cells (S) and the concomitant increase of enzyme activities a second increase of PAL activity was inducible by light. Actinomycin D or cycloheximide inhibited the "dilution effects" described above (HAHLBROCK and SCHRÖDER, 1975b).

Also in a later phase of subcultivation, the drastic increase of PAL activity was observed in cell cultures (S) from *Glycine max* (HAHLBROCK et al., 1971b) and from Paul's scarlet rose (DAVIES, 1972b). On the seventh day (*G. max*) and at the earliest on the third day (rose) of subcultivation, respectively, increased enzyme activity begins, followed by a complete decrease after some days. Cinnamic acid 4-hydroxylase and two isoenzymes of p-coumarate:CoA ligase in cell cultures (S) of *G. max* exhibit the same behavior as PAL. Illumination of the cells in this period may cause a remarkable quantitative change in the course of activities of the three enzymes (EBEL et al., 1974). In these experiments no pigment formation occurs. However, in an obvious correlation with the increase of PAL activity in cell cultures (S) of Paul's scarlet rose, the synthesis of polyphenols begins.

Additional arguments for the existence of coordinately regulated groups of enzymes involved in phenylpropane metabolism of cell cultures are given by the following results: During subcultivation of *Petunia hybrida* cell cultures (C) under continuous light, the course of activities of PAL, cinnamic acid 4-hydroxylase, and p-coumaric acid hydroxylase depends on the phytohormones present in the medium. Nevertheless, these group 1 enzymes show identical changes of activities under a specific condition. In contrast, chalcone-flavanone isomerase, which is a group 2 enzyme, and coniferyl alcohol dehydrogenase, an enzyme engaged in lignin biosynthesis, differ in their behavior from this group of enzymes and from each other (cf. Fig. 3 of LUCKNER and NOVER, p. 14). These findings indicate a third group of coordinately regulated enzymes (RANJEVA et al., 1975).

Summarizing this chapter, the following aspects are essential: The enzymes of light-dependent flavonoid synthesis are also detectable in the dark. In addition, despite the absence of light a remarkable increase of enzyme activities is possible (Fig. 49). The triggering mechanisms of these processes are unknown. However, the increase of activities in the dark is restricted to enzymes of the general phenylpropane metabolism (enzymes of group 1 in *P. hortense* cell cultures). Obviously, no substrate-induced formation of the enzymes engaged exclusively in flavonoid synthesis occurs in these cases. It appears that the decisive influence of light is directed to the formation of enzymes that

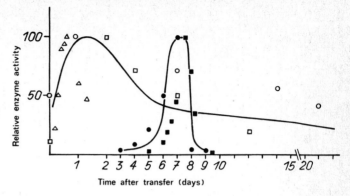

Fig. 49. Activities of enzymes - PAL, cinnamic acid 4-hydroxylase (CAH), p-coumarate:CoA ligase (CCoAL) - during subcultivation of plant cells in the dark. Lines indicate two types of maxima known from cultures of different species: ☐ *Citrus* species (PAL), ○ *H. gracilis* (PAL), △ *P. hortense* (PAL, CCoAL), ■ *G. max* (PAL, CAH, sum of two CCoAL isoenzymes), ● Paul's scarlet rose (PAL). (From THORPE et al., 1971; GREGOR and REINERT, 1972; HAHLBROCK and WELLMANN, 1973; HAHLBROCK et al., 1971b; EBEL et al., 1974; DAVIES, 1972b)

correspond to those of group 2 in cell cultures of *P. hortense*. - This detailed characterization of a differentiation process at the enzymic level should find general interest.

3. Comparison with the Related Intact Plant

While in the foregoing chapters individual secondary substances were neglected, in this part, spectra of secondary substances of cell cultures are compared with those of the related intact plants. Because some plant parts generally do not show biosynthetic activity or differ in their spectra of secondary substances from those of other parts, the basis of comparison must be clarified: Do cell cultures derived from various parts of the same plant differ with respect to their secondary metabolism?

Observations and experiments directed toward this question led to the finding that cell cultures from biosynthetically inactive tissues can, in principle, perform the plant-specific secondary metabolic processes. Recently, a clear example of this was reported by ZENK et al. (1975): In *M. citrifolia* plants, synthesis and storage of anthraquinones are restricted to the root. Nevertheless, tissues of stem, leaves, and fruits give rise to cell cultures (S) whose anthraquinone spectrum is identical to that of cell cultures (S) derived from root tissue. Although roots of *Papaver somniferum* do not show synthesis of alkaloids, cell cultures (C) from these plant parts as well as from capsules and stems of the poppy perform alkaloid metabolism (FURUYA et al., 1972). - The ability of cell cultures derived from biosynthetically inactive plant parts to synthesize secondary substances supports the earlier conclusion (cf. Chap. B) that the presence of secondary metabolism in the explants is no prerequisite for secondary metabolism in the originating callus.

Despite the obvious omnipotency of plant cells in respect to secondary metabolism, the following correlations between parts of *Ruta graveolens* and the derived cell cultures (C) occur: The volatile oil of intact plants consists of various aliphatic hydrocarbons (C9 and C11 compounds) and of some terpenes, the former predominating in the shoots, the latter in the roots. Cell cultures (C) from tissue of stems and leaves form almost exclusively C9 and C11 hydrocarbons under continuous light, while in the dark the terpenes predominate. In contrast, cell cultures (C) derived from root tissue synthesize exclusively terpenes in the light as well as in the dark. Obviously, their function is restricted to that of the root cells of intact plants (NAGEL and REINHARD, 1975).

There is no doubt that the different patterns of formation of volatile oil in cell cultures (C) of *R. graveolens* represents an exception. In general, cell cultures derived from plant parts that differ in their secondary metabolic reactions show a uniform secondary metabolism. Therefore, the origin of a cell culture may be neglected in this respect. - A qualitative comparison indicates that the ratio of secondary substances known from the plant is only seldom realized in the derived cell culture (e.g., anthocyanins in cell cultures (C) of *Dimorphotheca sinuata*: BALL et al., 1972). Frequently individual secondary substances of the spectra detectable in plants are absent in the related cell cultures. The cause of this fact need not be a change of differentiation processes leading to secondary metabolism under the conditions of cell cultures. It is reasonable to assume that in these systems individual reactions of secondary metabolism fail, for example, because of changed localization of substrate and enzyme. However, the general lack of some secondary substances requires attention. There is, for example, as yet no reliable detection of morphinanes in cell cultures of *Papaver* species (FURUYA et al., 1972; IKUTA et al., 1974) despite alkaloid metabolism performed by these systems.

On the other hand, secondary substances unknown in the intact plant may occur in the derived cell cultures. Apart from methodological difficulties incurred in clarifying this most remarkable fact, it is well documented by some investigators (VON BROCKE et al., 1971; STECK et al., 1971; see BALL et al., 1972). Even new natural compounds were isolated: some sesquiterpene lactones from cell cultures (C) of *Andrographis paniculata* (Fig. 50; BUTCHER and CONNOLLY, 1971) and norsanguinarine from cell cultures (C) of *P. somniferum* (FURUYA et al., 1972). In both cell cultures the typical terpene and alkaloid spectrum of the plants, respectively, was absent. With respect to these findings, it must be mentioned that in cell cultures, gene expressions occur that do not exist in the related intact plant.

In order to compare enzymatic reactions of secondary metabolism of intact plant tissues with those known from plant cell cultures, McCLURE and GROSS (1975) investigated the activities of PAL and hydroxycinnamate:CoA ligase after illumination of etiolated seedlings of five plant species including parsley. While PAL showed an increase of activity in each system, CoA ligase activity was practically unchanged. Obviously, in contrast to results obtained

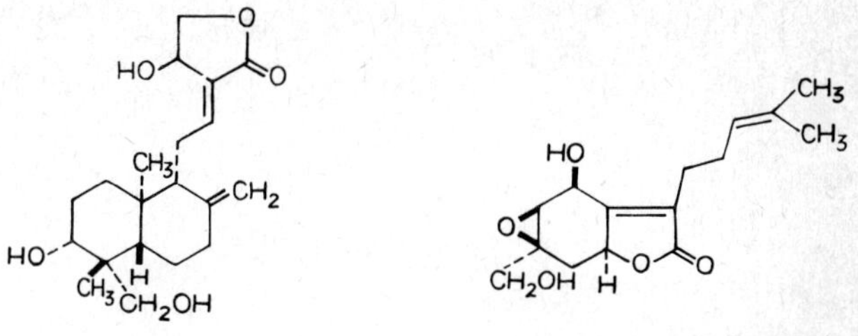

Andrographolide   Paniculide B

Fig. 50. Sesquiterpenes of paniculide B type are synthesized by cell cultures (C) from *Andrographis paniculata*. They are absent in related intact plants, which accumulate diterpenes of andrographolide type. (From BUTCHER and COLLONNY, 1971, redrawn)

with cell cultures (S) from *P. hortense* (cf. Chap. C2), these two enzymes are not coordinately regulated in young plants. - Apart from the necessity for continuing experiments in this field, the cited findings underline the general difficulty encountered in extrapolating the metabolic behavior of plant cell cultures to that of the related intact plants.

D. Correlation between Secondary Metabolism and Cellular Structures

According to electron microscopic studies of cell cultures (C) of *Pinus elliotti* (BAUR and WALKINSHAW, 1974), the smooth endoplasmic reticulum and Golgi bodies play a role in the synthesis of tannin or its precursors. At these membranous structures, tannin-containing cisternae and microvesicles, respectively, occur, which gradually enlarge and disengage. Large tannin deposits are always surrounded by at least one unit membrane. A short time after the occurrence of the first tannin deposite in the cytoplasm, an accumulation of tannin begins within the central vacuole. Finally this organelle contains the main amount of tannin and may fill the whole cell. Earlier studies with cell cultures (C) from *J. communis* also showed that in young tannin cells small tannin-containing vacuoles destributed through the cytoplasm predominate while mature cells store their tannin in the voluminous central vacuole. In the latter stage small optically empty vacuoles occur in the peripheral plasma layer, indicating a transfer of tannin from this vacuole type to the central vacuole (CONSTABEL, 1969).

Cells specialized in the synthesis and storage of tannin obviously do not differ morphologically from other cells. In contrast, cell cultures (T) of *R. graveolens* show typically formed spaces that store volatile oil (Fig. 51; REINHARD et al., 1968). Like in the intact plant, besides these so-called secretory cavities, ordinary cells occur containing volatile oil as droplets. The formation of the typical structures in cell cultures (T)

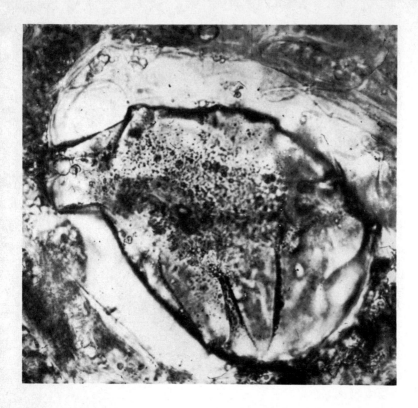

Fig. 51. Secretory cavity in cell cultures (C) from *R. graveolens* grown under continuous light, × 1500. (Photo was taken by REINHARD)

is promoted by light. In the dark the oil cells predominate. Because the composition of the volatile oil under these different cultural conditions is also different (cf. Chap. C3), one may conclude that synthesis and storage of volatile oil of a specific quality depend on a specific structure in cell cultures (T) from *R. graveolens* (REINHARD et al., 1968). Despite the influence of light, small groups of suspended *R. graveolens* cells do not form secretory cavities (REINHARD et al., 1971). Obviously, a specific state of organization is a prerequisite for this differentiation process.

The alkaloid-accumulating cells in cell cultures (T) of *Macleaya* species may also be briefly discussed here. They are scattered through the culture, and their morphology does not differ from that of ordinary cells. As a rule, they exhibit a yellow color caused by the alkaloid sanguinarine and a pigment without alkaloidal properties (NEUMANN and MÜLLER, 1967; KOBLITZ et al., 1975). An electron microscopic picture (Fig. 52) demonstrates the extensive vacuolization of these cells. With some reagents the alkaloids are detectable within the vacuoles (NEUMANN, unpubl. data). Probably, the alkaloid-accumulating cells of *Macleaya* cell cultures (T) represent latex cells like those found by KOHLENBACH (1965) in calli derived from leaf cells of *M. cordata*.

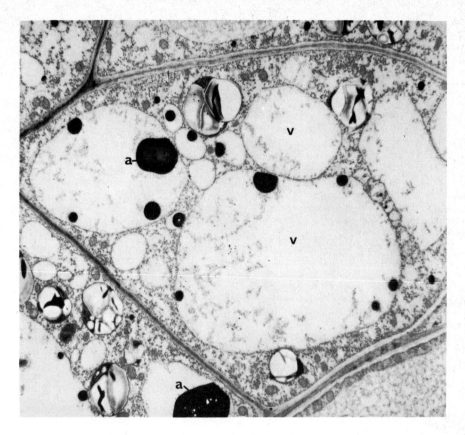

Fig. 52. Alkaloid-accumulating cell in cell cultures (C) from *M. cordata* grown under continuous light. v vacuole, a alkaloid precipitation; × 3600; glutaric aldehyde/hexachloroplatinic(IV) acid fixation. (Photo was taken by NEUMANN)

Mature *Macleaya* plants are characterized by latex vessels originating from several cells. Only leaf bud scales of *Macleaya* show single, colored latex cells (NEUMANN, unpubl. data). With respect to these properties, cell cultures from *Macleaya* species represent an early, relatively simple developmental stage of the intact plant.

Considering the cited results there is no doubt that a close correlation exists between secondary metabolism and cell structures, i.e., between biochemical and morphologic differentiation. However, it is unknown to what extent the realization of secondary metabolism depends on the formation of specific structures. In dark grown *P. hortense* cell cultures (S) the activity of the microsomal cinnamic acid 4-hydroxylase is increased about 4-fold after illumination. Under the same conditions cytidine-5'-diphosphocholine: 1,2-diacylglycerol cholinephosphotransferase shows a remarkable decrease of its original activity. The latter enzyme is involved in the biosynthesis of lecithin, a major phospholipid of the microsomal fraction of plant cells (SCHEEL and SANDERMANN, 1975).

E. Growth of Plant Cell Cultures and Formation of Secondary Substances

This chapter deals with the well-known problem of developmental physiology, i.e., whether growth and differentiation are compatible in the same biological system. It is related to a question discussed in chapter B concerning the ability of a differentiated cell to divide. A short description of pertinent results from cell cultures is possible from two points of view: (1) A more qualitative description, comparing the course of growth with that of formation of secondary substances during subcultivation and (2) a more quantitative approach in which the extents of both processes are determined. Usually the growth of cell cultures is determined by measuring their weights. The quantity of secondary products, expressed as the amount of substance per unit of biomass, may serve as the measure of differentiation. In this connection it must be considered that the fundamental steps of these differentiation processes may take place some time before the secondary substances are detected, although detailed information is lacking in this respect.

Ad 1: As is generally known, cell cultures of higher plants pass through several complex stages during subcultivation. After a lag phase cell division predominates followed by intensified cell growth. The latter stage normally causes a high growth rate of the culture. Final stages show a decrease of culture growth leading to a stationary phase.

Cell cultures (C) of *J. communis* that have been growing for about 30 days very slowly showed a maximum content of tannins at the end of this period (CONSTABEL, 1968). The highest density of tannin cells was found concomitantly (CONSTABEL, 1969). In the following phase of increased growth rate, the tannin content decreased and from 50 days after inoculation onwards remained constant despite the further increase of fresh weight. Other secondary substances in cell cultures that peak early are the alkaloids of *N. tabacum* (NEUMANN and MÜLLER, 1971). On the 5th day of the culture period, before increase of growth rate, the alkaloid content reaches a maximum and afterwards decreases continuously.

It has been more frequently reported that the content of secondary substances and the weight of cell culture reach their highest values at approximately the same time. This holds true also when the increased formation of secondary substances starts considerably later than that of culture growth. Data of such cases that are of special interest were given by CONSTABEL et al. (1971) and by DAVIES (1972a) concerning anthocyanin formation in cell cultures (S) of *H. gracilis* and polyphenol content in rose cell cultures (S), respectively. It was demonstrated that different concentrations of NAA, cause a more or less similar growth of cell cultures (S, *H. gracilis*), but different dynamics of anthocyanin formation (CONSTABEL et al., 1971).

Summarizing, one can say that obviously in a plant cell culture maxima of the content of secondary substances do not occur at the time of intensive cell division.

Ad 2: It is general experience that in cell cultures in different media a decreased growth rate is frequently related to an increased content of secondary substances. Examples include the flavone content in cell cultures (S) of *P. hortense* (HAHLBROCK and WELLMANN, 1970), the cyanidin content in *H. gracilis* cell cultures (S; FRITSCH et al., 1971), and the alkaloid content in cell cultures (S) of *N. tabacum* (NEUMANN and MÜLLER, 1971). Similarly the anthocyanin content of *D. carota* cell cultures (C) is much higher under irradiation with 7000 lux than under irradiation with 1000 lux. The former light intensity sharply inhibits the growth of the culture (ALFERMANN, 1973). In cell cultures (S) of *N. tabacum*, the growth of which is inhibited by 10^{-4} M D-threo-chloramphenicol, the alkaloid content at the end of the subcultivation was considerably higher than in control cultures. However, an addition of 10^{-6} M cycloheximide causes a sharp decrease of the growth rate but does not influence the alkaloid content (NEUMANN and MÜLLER, 1971).

It is essential to notice that there is no direct correlation between low growth rate and high content of secondary substances. This was emphasized also by ALFERMANN (1973): Dark grown cell cultures (C) of *D. carota* with or without 10^{-5} M 2,4-D have the same growth rate, although anthocyanins are synthesized only in the presence of 2,4-D. Evidently, in cell cultures, as in other biological systems, differentiation and growth are not mutually exclusive although there may be a mutual quantitative restriction.

F. Concluding Remarks

This chapter will stress once more the fact mentioned in the beginning of the paper, i.e., that cell cultures derived from plants exhibiting secondary metabolism frequently do not contain secondary substances. This situation is not only of scientific but also of practical interest, because cell cultures from medicinal plants might be valuable for the production of therapeutic substances. In this respect, possible reasons for the metabolic failure of cell cultures have been repeatedly discussed. An instructive compilation of these discussions was put together by TEUSCHER (1973). Such considerations result in the general supposition that specific differentiation processes are absent in cell cultures without secondary substances. But there is no detailed information about this problem. Thus, besides investigations of cell cultures performing secondary metabolism, experiments focused on selected systems unable to synthesize secondary substances may extend our knowledge of differentiation.

Acknowledgements. I thank Dr. H. KOBLITZ, Gatersleben, Dr. D. NEUMANN, Halle, Professor E. REINHARD, Tübingen, and Professor M.H. ZENK, Bochum, for showing me unpublished results and supplying me with unpublished photographs, respectively.

References

ALFERMANN, A.W.: Untersuchungen zur Anthocyansynthese in Calluskulturen von Daucus carota L. Dissertation: Eberhard-Karls-Universität Tübingen, 1973.

ALFERMANN, W., REINHARD, E.: Isolierung anthocyanhaltiger und anthocyanfreier Gewebestämme von Daucus carota: Einfluß von Auxinen auf die Anthocyanbildung. Experientia 27, 353-354 (1971).

BALL, E.A., HARBORNE, J.B., ARDITTI, J.: Anthocyanins of Dimorphotheca (Compositae) I. Identity of pigments in flowers, stems, and callus cultures. Am. J. Botany 59, 924-930 (1972).

BAUR, P.S., WALKINSHAW, C.H.: Fine structure of tannin accumulations in callus cultures of Pinus elliotti (slash pine). Can. J. Botany 52, 615-619 (1974).

BHATT, P.H., MEHTA, A.R.: Growth and differentiation in mechanically isolated mesophyll cells of Ipomoea quamoclit. Can. J. Botany 52, 217-218 (1974).

BILLY, F. DE, PAUPARDIN, C.: Sur l'évolution des huiles essentielles dans les tissus des péricarpe de Citron (Citrus Limonia Obseck) cultivés in vitro. C. R. Séanc. Acad. Sci., Serie D, 273, 1690-1693 (1971).

BROCKE, W. VON, REINHARD, E., NICHOLSON, G., KÖNIG, W.A.: Über das Vorkommen von 3-(1',1'-Dimethylallyl)-scopoletin in Gewebekulturen von Ruta graveolens. Z. Naturforsch. 26b, 1252-1255 (1971).

BRUNET, G., IBRAHIM, R.K.: Tissue culture of Citrus peel and its potential for flavonoid synthesis. Z. Pflanzenphysiol. 69, 152-162 (1973).

BUTCHER, D.N., CONNOLLY, J.D.: An investigation of factors which influence the production of abnormal terpenoids by callus cultures of Andrographis paniculata Nees. J. Exp. Botany 22, 314-322 (1971).

CONSTABEL, F.: Gerbstoffproduktion der Calluskulturen von Juniperus communis L. Planta (Berlin) 79, 58-64 (1968).

CONSTABEL, F.: Über die Entwicklung von Gerbstoffzellen in Calluskulturen von Juniperus communis L. Planta Med. (Stuttgart) 17, 101-115 (1969).

CONSTABEL, F., SHYLUK, J.P., GAMBORG, O.L.: The effect of hormones on anthocyanin accumulation in cell cultures of Haplopappus gracilis. Planta (Berlin) 96, 306-316 (1971).

DAVIES, M.E.: Polyphenol synthesis in cell suspension cultures of Paul's scarlet rose. Planta (Berlin) 104, 50-65 (1972a).

DAVIES, M.E.: Effects of auxin on polyphenol accumulation and the development of phenylalanine ammonia-lyase activity in dark grown suspension cultures of Paul's scarlet rose. Planta (Berlin) 104, 66-77 (1972b).

EBEL, J., SCHALLER-HEKELER, B., KNOBLOCH, K.-H., WELLMANN, E., GRISEBACH, H., HAHLBROCK, K.: Coordinated changes in enzyme activities of phenylpropanoid metabolism during the growth of soybean cell suspension cultures. Biochim. Biophys. Acta 362, 417-424 (1974).

FRITSCH, H., HAHLBROCK, K., GRISEBACH, H.: Biosynthese von Cyanidin in Zellsuspensionskulturen von Haplopappus gracilis. Z. Naturforsch. 26b, 581-585 (1971).

FURUYA, T., IKUTA, A., SYONO, K.: Alkaloids from callus tissue of Papaver somniferum. Phytochemistry 11, 3041-3044 (1972).
FURUYA, T., KOJIMA, H., SYONO, K.: Regulation of nicotine synthesis in tobacco callus tissue. Chem. Pharmac. Bull. (Tokyo) 15, 901-903 (1967).
FURUYA, T., KOJIMA, H., SYONO, K.: Regulation of nicotine biosynthesis by auxins in tobacco callus tissues. Phytochemistry 10, 1529-1532 (1971).
GREGOR, H.-D., REINERT, J.: Induktion der Phenylalanin-Ammonium-Lyase in Gewebekulturen von Haplopappus gracilis. Protoplasma 74, 307-319 (1972).
GRISEBACH, H.: Biosynthesis of flavonoids. In: Chemistry and Biochemistry of Plant Pigments. GOODWIN, T.W. (ed.). New York-London: Academic Press 1965.
GRISEBACH, H., BARZ, W.: Biochemie der Flavonoide. Naturwissenschaften 56, 538-544 (1969).
HAHLBROCK, K., EBEL, J., ORTMANN, R., SUTTER, A., WELLMANN, E., GRISEBACH, H.: Regulation of enzyme activities related to the biosynthesis of flavone glycosides in cell suspension cultures of parsley (Petroselinum hortense). Biochim. Biophys. Acta 244, 7-15 (1971a).
HAHLBROCK, K., KNOBLOCH, K.-H., KREUZALER, F., POTTS, J.R.M., WELLMANN, E.: Coordinated induction and subsequent activity changes of two groups of metabolically interrelated enzymes. Light-induced synthesis of flavonoid glycosides in cell suspension cultures of Petroselinum hortense. Europ. J. Biochem. 61, 199-206 (1976).
HAHLBROCK, K., KUHLEN, E., LINDL, T.: Änderungen von Enzymaktivitäten während des Wachstums von Zellsuspensionskulturen von Glycine max: Phenylalanin Ammonium-Lyase und p-Cumarat:Co A Ligase. Planta (Berlin) 99, 311-318 (1971b).
HAHLBROCK, K., RAGG, H.: Light-induced changes of enzyme activities in parsley cell suspension cultures. Effects of inhibitors of RNA and protein synthesis. Arch. Biochim. Biophys. 166, 41-46 (1975).
HAHLBROCK, K., SCHRÖDER, J.: Light-induced changes of enzyme activities in parsley cell suspension cultures. Increased rate of synthesis of phenylalanine ammonia-lyase. Arch. Biochem. Biophys. 166, 47-53 (1975a).
HAHLBROCK, K., SCHRÖDER, J.: Specific effects on enzyme activities upon dilution of Petroselinum hortense cell cultures into water. Arch. Biochem. Biophys. 171, 500-506 (1975b).
HAHLBROCK, K., WELLMANN, E.: Light-induced flavone biosynthesis and activity of phenylalanine ammonia-lyase and UDP-apiose synthetase in cell suspension cultures of Petroselinum hortense. Planta (Berlin) 94, 236-239 (1970).
HAHLBROCK, K., WELLMANN, E.: Light-independent induction of enzymes related to phenylpropanoid metabolism in cell suspension cultures from parsley. Biochim. Biophys. Acta 304, 702-706 (1973).
HEBLE, M.R., NARAYANASWAMY, S., CHADHA, M.S.: Tissue differentiation and plumbagin synthesis in variant cell strains of Plumbago zeylanica L. in vitro. Plant Sci. Letters 2, 405-409 (1974).
HESS, D., ENDRESS, R.: Anthocyansynthese in isolierten Protoplasten von Nemesia strumosa var. Feuerkönig. Z. Pflanzenphysiol. 68, 441-449 (1973).

IBRAHIM, R.K., THAKUR, M.L., PERMANAND, B.: Formation of anthocyanins in callus tissue cultures. Lloydia 34, 175-182 (1971).
IKUTA, A., SYONO, K., FURUYA, T.: Alkaloids of callus tissues and redifferentiated plantlets in the Papaveraceae. Phytochemistry 13, 2175-2179 (1974).
JENSEN, R.G., FRANCKI, R.J.B., ZAITLIN, M.: Metabolism of separated leaf cells. I. Preparation of photosynthetically active cells from tobacco. Plant Physiol. 48, 9-13 (1971).
KOBLITZ, H., SCHUMANN, U., BÖHM, H., FRANKE, J.: Gewebekulturen aus Alkaloidpflanzen IV. Macleaya microcarpa (Maxim.) Fedde. Experientia 31, 768-769 (1975).
KOHLENBACH, H.W.: Die Entwicklungspotenzen explantierter und isolierter Dauerzellen II. Das zelluläre Differenzierungswachstum bei Blattzellkulturen von Macleaya cordata. Beitr. Biol. Pflanz. 41, 469-480 (1965).
KREUZALER, F., HAHLBROCK, K.: Flavonoid glycosides from illuminated cell suspension cultures of Petroselinum hortense. Phytochemistry 12, 1149-1152 (1973).
KRIKORIAN, A.D., STEWARD, F.C.: Biochemical differentiation: The biosynthetic potentialities of growing and quiescent tissue. Plant Physiol. V B, 227-326 (1969).
LACKMANN, I.: Wirkungsspektren der Anthocyansynthese in Gewebekulturen und Keimlingen von Haplopappus gracilis. Planta (Berlin) 98, 258-269 (1971).
McCLURE, J.M., GROSS, G.G.: Diverse photoinduction characteristics of hydroxycinnamate:coenzyme A ligase and phenylalanine ammonia lyase in dicotyledonous seedlings. Z. Pflanzenphysiol. 76, 51-55 (1975).
NAGEL, M., REINHARD, E.: Das ätherische Öl der Calluskulturen von Ruta graveolens II. Physiologie zur Bildung des ätherischen Öles. Planta Med. (Stuttgart) 27, 264-271 (1975).
NEUMANN, D., MÜLLER, E.: Intrazellulärer Nachweis von Alkaloiden in Pflanzenzellen im licht- und elektronenmikroskopischen Maßstab. Flora, Sect. A, 158, 479-491 (1967).
NEUMANN, D., MÜLLER, E.: Beiträge zur Physiologie der Alkaloide V. Alkaloidbildung in Kallus- und Suspensionskulturen von Nicotiana tabacum L. Biochem. Physiol. Pflanz. 162, 503-513 (1971).
PELCHER, L.E., GAMBORG, O.L., KAO, K.N.: Bean mesophyll protoplasts: Production, culture and callus formation. Plant Sci. Letters 3, 107-111 (1974).
RANJEVA, R., BOUDET, A.M., HARADA, H., MARIGO, G.: Phenolic metabolism in Petunia tissues I. Characteristic responses of enzymes involved in different steps of polyphenol synthesis to different hormonal influences. Biochim. Biophys. Acta 399, 23-30 (1975).
REINERT, J., CLAUSS, H., ARDENNE, R. VON: Anthocyanbildung in Gewebekulturen von Haplopappus gracilis in Licht verschiedener Qualität. Naturwissenschaften 51, 87 (1964).
REINHARD, E., CORDUAN, G., BROCKE, W. VON: Untersuchungen über das ätherische Öl und die Cumarine in Gewebekulturen von Ruta graveolens. Herba Hung. 10, 9-26 (1971).
REINHARD, E., CORDUAN, G., VOLK, O.H.: Über Gewebekulturen von Ruta graveolens. Planta Med. (Stuttgart) 16, 8-16 (1968).
SCHEEL, D., SANDERMANN, H.: On the mechanism of light induction of plant microsomal cinnamic acid 4-hydroxylase. Planta (Berlin) 124, 211-214 (1975).

SCHMITZ, M., SEITZ, U.: Hemmung der Anthocyansynthese durch Gibberellinsäure A_3 bei Kalluskulturen von Daucus carota. Z. Pflanzenphysiol. 68, 259-265 (1972).
STECK, W., BAILEY, B.K., SHYLUK, J.P., GAMBORG, O.L.: Coumarins and alkaloids from cell cultures of Ruta graveolens. Phytochemistry 10, 191-194 (1971).
STREET, H.E. (ed.): Plant Tissue and Cell Culture. Blackwell Scientific Publications, Oxford 1973.
SUTTER, A., GRISEBACH, H.: UDP-Glucose: flavonol 3-O-glucosyltransferase from cell suspension cultures of parsley. Biochim. Biophys. Acta 309, 289-295 (1973).
TABATA, M., MIZUKAMI, H., HIRAOKA, N., KONOSHIMA, M.: Pigment formation in callus cultures of Lithospermum erythrorhizon. Phytochemistry 13, 927-932 (1974).
TEUSCHER, E.: Probleme der Produktion sekundärer Pflanzenstoffe mit Hilfe von Zellkulturen. Pharmazie 28, 6-18 (1973).
THORPE, T.A., MAIER, V.P., HASEGAWA, S.: Phenylalanine ammonialyase activity in citrus fruit tissue cultured in vitro. Phytochemistry 10, 711-718 (1971).
WELLMANN, E.: Phytochrome-mediated flavone glycoside synthesis in cell suspension cultures of Petroselinum hortense after preirradiation with ultraviolet light. Planta (Berlin) 101, 283-286 (1971).
WELLMANN, E.: UV dose-dependent induction of enzymes related to flavonoid biosynthesis in cell suspension cultures of parsley. FEBS Letters 51, 105-107 (1975).
WELLMANN, E., BARON, D.: Durch Phytochrom kontrollierte Enzyme der Flavonoidsynthese in Zellsuspensionskulturen von Petersilie (Petroselinum hortense Hoffm.). Planta (Berlin) 119, 161-164 (1974).
WELLMANN, E., SCHOPFER, P.: Phytochrome-mediated de novo synthesis of phenylalanine ammonia-lyase in cell suspension cultures of parsley. Plant Physiol. 55, 822-827 (1975).
ZENK, M.H., EL-SHAGI, H., SCHULTE, U.: Anthraquinone production by cell suspension cultures of Morinda citrifolia. Planta Med. (Stuttgart) Suppl. 1975, 79-101.

Subject Index

Abbreviations: (C), (T), (S): Plant cell cultures, E: Action as effector, R: Regulation of synthesis, amount, activity, degradation etc.

Accumulation of secondary products 105-107
Acetylcholine, E: 20; R: 9
 esterase, R: 20
ACTH → Adrenocorticotrophic hormone
Actinomyces streptomycini 18, 22-24
Actinomycin, R: 8; → Inhibitors of gene expression
Activation of phenylalanine ammonia-lyase 31
Adaption, enzymatic 37
Adenotropic hormone, E: 78
Adenyl cyclase, R: 33
Adrenal cortex 33-36
Adrenocorticotrophic hormones, E: 20, 33-36
A-factor, E: 18, 22-24
Alkaloids, R: 8, 13-22, 39-58, 106, 109, 113, 114, 116, 118, 119
Alpinigenine, R: 13-16
α-Amanitin → Inhibitors of gene expression
Amaranthus spec. 9, 19, 27
Amino acids, level after cycloheximide action 50, 51
Andrographis paniculata (C): 114
Andrographolide, R: 115; → Terpenes
Animal hormones → Hormones
Animals, secondary metabolism, R: 9, 20, 33-36, 74-76, 77-79
Anther development 72-74
Anthocyanins, biosynthetic pathway 12; R: 24-27, 73, 74, 106-109, 114, 118, 119; → Flavonoids
Anthraquinones, R: 20, 109, 113
Apiin, biosynthetic pathway 12
Apiosyltransferase, R: 11-14, 27, 110
Aspergillus fumigatus 8, 18
Auxins, E: 19, 20, 109, 118, 119; → Phytohormones

Bacillus spec. 8, 58-63
Bacterial sporulation 58-63
Benzodiazepine alkaloids, R: 8, 18, 39-58; → Alkaloids
Benzoic acid derivatives, biosynthetic pathway 11
 formation in chloroplasts 71
 E: 31

Betacyanins → Betalains
Betalains, R: 9, 19, 27
Biosynthesis of secondary products 105-107, 113, 118, 119
Blakeslea trispora 8, 18, 19, 65-70
Blowfly → *Calliphora erythrocephala*
Brassica oleracea 13, 27
Buckwheat → *Fagopyron esculentum*

Caffeic acid, biosynthetic pathway 11
 E: 25; → Cinnamic acid derivatives
Calliphora erythrocephala 9, 20, 77-79
cAMP → Cyclic adenosine monophosphate
Carbon dioxide, E: 21
Carotenoids, R: 8, 18-20, 27, 66-70, 72-73
Catabolite repression → Glucose
Cell cultures → Plant cell cultures
Cell division, formation of secondary enzymes 44
 formation of specialization proteins 55; → Division of specialized cells
Cell specialization, general 5, 37, 105
Celosia plumosa 9, 19, 27
Chalcone-flavanone isomerase, R: 11-14, 27, 73, 74, 110, 112
Chalcones, biosynthetic pathway 12
 R: 73, 74; → Flavonoids
Channelling, in cinnamic acid metabolism 71; → Compartmentation
Chanoclavine cyclase, R: 21
Chemical specialization → Cell specialization
Chironomus tentans 79
Chloramphenicol → Inhibitors of gene expression
2-(4-Chlorophenylthio)-triethylamine, E: 18, 20
Chloroplasts 70, 71
Cholesterol, cleavage of side chain 33
Choline acetyltransferase, R: 20
Chromatin, function of 38
Cinnamate: CoA ligase, R: 11-14; → p-Coumarate: CoA ligase

Cinnamate 4-hydroxylase, R: 11-14, 27, 110, 112, 117
Cinnamic acid derivatives, biosynthetic pathway 11
  E: 19, 24-26, 31
  R: 8, 11-14, 19, 27, 73, 74
Citrus paradisi 20
Citrus spec., pericarp 107
  (T): 111
Claviceps spec. 8, 17-22
Compartmentation, of cholesterol in adrenal cortex 36
  of alkaloid precursors in *Penicillium cyclopium* 43
  of cyclopenase and its substrates 46
  of effectors 17
  of phytochrome 28
  of flavonoids in chloroplasts 70
  of phenylpropanoids in chloroplasts 70
  during pollen formation 72
Competence, general 6, 39
  and effector action 17
  in ergolin biosynthesis 21
  in streptomycin biosynthesis 23, 24
  in anthocyanin biosynthesis 26
  in phytochrome action 31-33
  in benzodiazepine alkaloid biosynthesis 57
  in urea biosynthesis 75
  in sclerotization and melanization of *Calliphora larvae* 77
Conidiation of *Penicillium cyclopium* 39-41, 44-47, 53-58
Coniferyl alcohol dehydrogenase, catalyzed reaction 11
  R: 13, 14, 112
Coordinate enzyme formation, general 10
  in flavonoid biosynthesis 11-14, 110, 115
  in lignin biosynthesis 13, 14, 112
  in ergolin alkaloid biosynthesis 21
  in benzodiazepine alkaloid biosynthesis 42, 44, 45
  in patulin biosynthesis 65
  in urea biosynthesis 75
Coordination of differentiation processes 39
Cortisol → Corticosteroids
Corticosteroids, R: 9, 20, 33-36
p-Coumarate: CoA ligase, R: 11-14, 27, 73, 74, 110, 111, 114
  isoenzymes 112
p-Coumarate hydroxylase, R: 11-14, 112
p-Coumaric acid, biosynthetic pathway 11
  E: 25; → Cinnamic acid derivatives

Cucumis sativus 31
Cucurbita moschata 20
Cyanidin, R: 25, 119; → Anthocyanins
Cyclic adenosine monophosphate, E: 5, 16, 19-21
  as second messenger 33-36
  and catabolite repression 53, 54
  action in *Penicillium cyclopium* 54
  action in *Penicillium urticae* 65
Cycloheximide → Inhibitors of gene expression; side effects 31, 50
Cyclopenase, catalyzed reaction 41
  R: 44-47
Cyclopenin → Benzodiazepine alkaloids
Cyclopenol → Benzodiazepine alkaloids
Cyclopeptine → Benzodiazepine alkaloids
— dehydrogenase, R: 41-45, 48-50, 57
Cytidine-5'-diphosphocholine: 1,2-diacylglycerol cholinephosphotransferase, R: 117
Cytokinins, E: 19; → Phytohormones

Darkness, E: 108, 110-112, 114, 116, 119
Daucus carota (C), (T): 19, 109, 119
Dehydrocyclopeptine → Benzodiazepine alkaloids
Dehydrocyclopeptine epoxidase, R: 41-45, 48
Delphinidin, R: 25; → Anthocyanins
Despecialization 106, 107
Determination, general 6, 39
  in ergolin biosynthesis 21
  in streptomycin biosynthesis 23
  in benzodiazepine alkaloid biosynthesis 55
  after phytochrome action 31
Developmental program, general 4, 39
Differentiation, general concept 3-6, 105
Differential gene expression, general 4
  in vitro investigation 6
Differentiated state 106, 107
Differentiation program, general 4, 37-39
  of the hyphae and conidiospores of *Penicillium cyclopium* 39-58
  of bacterial sporulation 58-63
  of the hyphae of *Penicillium urticae* 63-65
  of the cells of *Mucoraceae* 65-70
  of pollen cells 72-74
  of the liver cells of *Rana catesbeiana* 74-76
  of epidermis cells of *Calliphora larvae* 77-79

11β,21-Dihydroxy-4,17(20)-pregnadien-
  3-on, E: 18
Dimorphotheca sinuata (C): 114
Dipicolinic acid, R: 8, 60-62
Division of specialized cells 46,
  106, 107, 118; → Cell division
DMAT-Synthetase, R: 21
DNA, sequencing 6; → Inhibitors
  of DNA synthesis
Dopa, R: 9, 77-79
Dopadecarboxylase, R: 20, 77-79
  in vitro synthesis 79
Dunaliella marina 70

Ecdysone, E: 20, 77-79; → Hormones
Effectors → Regulatory effectors
— group specific → Nonsubstrate-
  like effectors
Enzymatic adaption 37
Enzyme activity, R: 31, 42-44, 110-
  113
— degradation 31, 80
Enzymes → Proteins
Ergoline alkaloids, R: 8, 17, 19,
  21, 22; → Alkaloids
Ethidium bromide → Inhibitors of
  DNA synthesis
Ethylene, E: 19
  R: 9, 19, 27

Fagopyrom esculentum 13, 28
Ferulic acid, biosynthetic pathway
  11
  E: 25, 26; → Cinnamic acid deriva-
    tives
Flavanone glycosides → Flavonoids
— synthetase, R: 12, 100
Flavone glycosides, R: 108-110, 119;
  → Flavonoids
Flavonoids, E: 31
  R: 11-14, 24-26, 27-33, 73, 74,
    105-110, 112, 118, 119
  in chloroplasts 70
Flavonole glycosides, R: 108; →
  Flavonoids
p-Fluorophenylalanine → Inhibitors
  of gene expression
5-Fluorouracil → Inhibitors of gene
  expression

β-Galactosidase, R: 48, 50, 52
Gene activation 4
Genes, in vitro synthesis and tran-
  scription 6
  visualization 6
Genetic varieties → Mutants
Gentisaldehyde, R: 63, 64
Gentisyl alcohol, R: 63, 64
Gherkin → *Cucumis sativus*
Gibberellic acid, E: 19
  R: 27
Glucose, E: 53, 54, 62, 65

Glucosyltransferase, R: 11-14, 27,
  110
Glycine max (S): 112
Gramicidin → Peptide antibiotics
Growth of plant cell cultures, R:
  118, 119

Haplopappus gracilis (S), (T): 106,
  108, 111, 112, 118, 119
Hormone receptor 34, 79
Hormones, E: 5, 16, 19, 20, 24, 33,
  65, 75, 77, 78, 109, 112, 118,
  119
  as secondary products 3
Hydrocarbons, aliphatic, R: 114; →
  Volatile oils
m-Hydroxybenzyl alcohol dehydrogenase, R: 64, 65
4-Hydroxycinnamate 3-hydroxylase →
  p-Coumarate hydroxylase
5-Hydroxyferulic acid, E: 25; →
  Cinnamic acid derivatives
20β-Hydroxysteroid dehydrogenase,
  R: 18
Hydroxyurea → Inhibitors of DNA
  synthesis

Idiophase, general 37
  of *Claviceps spec.* 21, 22
  of *Penicillium cyclopium* 39-58
  of bacteria 58-63
  of *Penicillium urticae* 63-65
  of *Mucoraceae* 66-70
  during anther development 72-74
Immunoprecipitation of proteins →
  Proteins
Immunocytochemistry and phytochrome
  distribution 28
Inactivation of phenylalanine ammonia-lyase 31
Inhibitors, of DNA synthesis 44, 46
  of gene expression 7, 21, 26, 29,
    35, 36, 46-53, 59, 61, 65, 67, 77,
    79, 110, 112, 119
Invertase, R: 44
Irradiation → Light
Isorhamnetin → Flavonoids

β-Jonone, E: 18
Juniperus communis, secondary phloem
  107
  (C): 115, 118
Juvenile hormone, E: 77

Kaempferol → Flavonoids

Lag phase, after signal reception,
  general 29
  after phytochrome action 29
  after ACTH action 35
  after ecdysone action 79
  of protein translation 36

Lecithin, R: 117
Light, E: white 11, 108-112, 114,
  116, 117, 119
  blue   108, 111, 112
  red   108, 109; → Phytochrome
  UV   109
Lignin, biosynthetic pathway 11
  R: 13, 14, 19, 112
Lilium henryi 72
Lipoxygenase, R: 32
Lithospermum erythrorhizon (C):
  109
Liver, chemical specialization
  74-76
Lycopersicon esculentum 20, 27

Macleaya spec. 116, 117
  (S), (T): 106, 116, 117
Malus silvestris 9
Mandelic acid, R: 63
Melanin, R: 45, 77
Messenger ribonucleic acids, processing of precursors 4
  stable species 36
  in vitro translation 6, 30, 79
Metabolic excretion 3
6-Methylsalicylic acid, R: 8, 63-65,
  70, 71
  decarboxylase, R: 64, 65
  synthetase complex, R: 64, 65
Microbial hormones → Hormones
Microorganisms, secondary metabolism, R: 8, 17-24, 39-58, 58-63,
  63-65, 65-70
Microsporogenesis 72-74
Morinda citrifolia (S): 20, 109, 113
Morphinanes, R: 114; → Alkaloids
Morphological specialization → Cell specialization
mRNA → Messenger ribonucleic acid
Mucor spec. 19, 65-70
Mutants (genetic varieties):
  Papaver bracteatum 13-16
  Actinomyces streptomycini 22-24
  Penicillium cyclopium 55-58
  Bacillus spec. 61, 62
  Mucoraceae 69, 70

Naphthoquinones, R: 109
Neoblastoma tumor cells 9, 20
Nicotiana tabacum (C): 109, 118, 119
Noncoordinate enzyme formation,
  general 10
  in alpinigenine biosynthese 13-16
  of UDP-apiose synthetase 13, 14
  of cyclopenase 44-47
  of 6-methyl salicylic acid synthetase 63-65
Nonhistone proteins, function of 38
Nonsubstrate-like effectors 4, 5,
  10, 16, 18-20, 22-24, 27-33, 33-36,
  53, 54, 56-58, 74-76, 77-79, 109,
  112, 118, 119
Norsanguinarine, R: 114; → Alkaloids

Orsellinic acid, R: 8

PAL → Phenylalanine ammonia-lyase
Paniculide B, R: 115; → Terpenes
Papaver bracteatum 13
  spec. (C): 113, 114
Papaverine, E: 20
Paul's scarlet rose (S): 112, 118
Parsley → *Petroselinum hortense*
Patulin, R: 8, 63-65
Penicillium cyclopium 8, 18, 39-58
— *urticae* 8, 63-65
Peonidin, R: 25, 26; → Anthocyanins
Peptide antibiotics, E: 62, 63
  R: 8, 60-62
— hormones → Adrenocorticotrophic
  hormone; → Adenotrophic hormones
Peroxidase, R: 72
Petroselinum hortense 13, 14
  (S): 11, 30, 108-112, 114, 117,
  119
Petunia hybrida 19, 24-26
  (C): 13, 14, 112
Petunidin, R: 25; → Anthocyanins
Phase dependence of secondary metabolism, general 37-39; → Differentiation programs; → Idiophase;
  → Trophophase
Phaseollin, R: 9
Phaseolus vulgaris 9
Phenolase → 4-Hydroxycinnamate
  3-hydroxylase
Phenol oxidase, R: 20, 44, 78
Phenylalanine ammonia-lyase, R:
  11-14, 27-32, 73, 74, 110-112,
  114
  de novo synthesis 8, 30, 31, 110
  turnover 31
  in vitro synthesis 30
  in chloroplasts 30
— analogues, E: 18, 58
Phenylpropanoid metabolism, R:
  11-14, 24-26, 27-33, 73, 74, 105-
  110, 112, 118, 119
  in chloroplasts 70, 71; → Cinnamic
  acid derivatives, → Flavonoids
Phycomyces spec. 19, 66
Phytochrome, E: 16, 19, 27-33
  R: 109
  distribution 28
  changes in response system 31-33
  in chloroplast development 70, 71
Phytohormones, E: 19, 20, 109, 112,
  118, 119; → Hormones
Phytosterols, R: 109; → Terpenes
Pisum sativum 9
Pinus elliotti (C): 115

Plant cell cultures, secondary metabolism, R: 9, 11-14, 19, 20, 30, 105ff.
  terminology 105
Plants, secondary metabolism, R: 8, 13, 14, 19, 20, 24-26, 27-33, 70, 71, 72-74
Plumbago zeylanica (C): 109
Pollen formation 72-74
Polyketides, R: 8, 20, 63-65, 70, 71; → Flavonoids
Polyphenols, R: 112, 118, 119
Polysomes, quantification 6
Pool → Compartmentation
Primary metabolism, definition 3
Processing, of messenger ribonucleic acid precursors 4
  of proteinogens 4, 46, 79
Proteinogen processing 4, 46, 79
Programs of development → Developmental programs
  — of differentiation → Differentiation programs
Prostaglandins, E: 35
Proteins, isotope-labeling 7, 29-31
  quantification by immunologic precipitation 7, 30, 77, 79
Protein kinases, R: 33
Protoplasts from *Nemesia strumosa* 106
Puromycin → Inhibitors of gene expression

Quantal cell cycle during conidiation of *Penicillium cyclopium* 46
Quercetin → Flavonoids

Radish → *Raphanus sativus*
Rana catesbeiana 9, 20, 74-76
Raphanus sativus 13
Receptor proteins 38
Red cabbage → *Brassica oleracea*
Regulatory effectors, phase dependence of activity 10
  action on programs 38;
  → nonsubstrate-like effectors;
  → substrate-like effectors
  — proteins 4-6, 31, 38, 52
  — ribonucleic acids 4, 38, 52
  — units → Coordinate enzyme formation
Ribonucleic acid sequencing 6
Rhodotorula spec. 18
RNA → Ribonucleic acids
RNA polymerase, proteolytic modification 62
Ruta graveolens (C), (S): 114-116

Sclerotization of insect larvae 77-79
Secondary metabolism, definition 3, 4

products, function 3
Sequential gene expression,
  in *Pseudomonas spec.* 63
  in *Penicillium urticae* 63-65
  in *Mucoraceae* 65-70
  in microsporogenesis 72-74
Sex hormones, of *Mucoraceae* 65-70
  → Hormones
Signals, reception and transformation 6, 16, 17; → Regulatory effectors
Sinapic acid, E: 25; → Cinnamic acid derivatives
Sinapis alba 13, 29-33
Sorghum vulgare 19
Specialization growth → Cell specialization
  — proteins → Cell specialization
Specialized cells in plant cell cultures 106, 107, 115-118
Spectra of secondary products 109, 113, 114
Spore pigments, R: 44, 60
Sporopollenin, R: 19, 68, 72
Sporulation, of bacteria 58-63
  of *Penicillium cyclopium* 39-41, 44-47, 53-58
Steroid carrier protein 36
  — hormones, E: 20, 38, 77-79;
  → Hormones
Steroids, E: 20, 38, 77-79
  R: 8, 9, 20, 33-36
Storage, of secondary products: 105, 106
  of alkaloids 106, 116
  of anthraquinones 113
  of tannins 115
  of volatile oils 107, 115, 116
Streptomyces spec. 8, 18
Streptomycin, R: 18, 22-24
Subcellular structures 107, 115-117
Subcultivation of plant cell cultures, stages of 108, 110, 118
Substrate-like effectors 4, 16, 22, 24-26, 31, 56-58, 63-65, 65-70
Sulfolactic acid, R: 8, 60, 61
Superinduction 51

Tanning agents in cuticula of insects, R: 9, 77-79
Tannins, R: 105, 107, 115, 118
Terpenes, R: 109, 114, 115; → Volatile oils
Thebaine, R: 13-16
β-Thienylalanine → Inhibitors of gene expression
Thyroid hormones, E: 20, 74-76;
  → Hormones
Triggering factor in plant cell cultures, definition of 108; →
  Regulatory effectors

Transcription, general 4
  in vitro 7
Transcriptional control during bacterial sporulation 59
Translation, general 4
  in vitro 7, 30, 79
Translational control during bacterial sporulation 59
Trijodothyronine → Thyroid hormones
Trisporic acids, E: 19, 65-70
  R: 8, 65-70
Triticum aestivum 27
Trophophase, general 37; → Idiophase
Tryptophan and analogues, E: 17-22
Tulipa spec. 73
Turnover, of phenylalanine ammonia-lyase 31
  of the rate limiting protein of corticosteroid formation 36

Tyrocidine → Peptide antibiotics
Tyrosine hydroxylase, R: 20
Tyrothricin → Peptide antibiotics

UDP-apiose synthetase, R: 11-14, 27, 110
Urea 9, 74-76
— cycle enzymes, R: 74-76

Viridicatin, biosynthetic pathway 41
  R: 44-47
Viridicatol, biosynthetic pathway 41
  R: 44-47
Virus formation 37
Volatile oils, R: 107, 114-116

White mustard → *Sinapis alba*

Zygospores of *Mucoraceae* 65-70

Molecular Biology, Biochemistry and Biophysics
Editors: A. Kleinzeller, G.F. Springer, H.G. Wittmann

Vol.1: J.H. van't Hoff
Imagination in Science
Translated into English with notes
and a general introduction by G.F. Springer
1 portrait. VI, 18 pp. 1967
ISBN 3–540–03933–3

Vol. 2: K. Freudenberg, A.C. Neish
Constitution and Biosynthesis of Lignin
10 figs. IX, 129 pp. 1968
ISBN 3–540–04274–1

Vol. 3: T. Robinson
The Biochemistry of Alkaloids
37 figs. X, 149 pp. 1968
ISBN 3–540–04275–X

Vol. 5: B. Jirgensons
Optical Activity of Proteins and Other Macromolecules
2nd revised and enlarged edition
71 figs. IX, 199 pp. 1973
ISBN 3–540–06340–4

Vol. 6: F. Egami, K. Nakamura
Microbial Ribonucleases
5 figs. IX, 90 pp. 1969
ISBN 3–540–04657–7

Vol. 8: **Protein Sequence Determination**
A Sourcebook of Methods and Techniques
Edited by S.B. Needleman
2nd revised and enlarged edition
80 figs. XVIII, 393 pp. 1975
ISBN 3–540–07256–X

Vol. 9: R. Grubb
The Genetic Markers of Human Immunoglobulins
8 figs. XII, 152 pp. 1970
ISBN 3–540–05211–9

Vol. 10: R.J. Lukens
Chemistry of Fungicidal Action
8 figs. XIII, 136 pp. 1971
ISBN 3–540–05405–7

Vol. 11: P. Reeves
The Bacteriocins
9 figs. XI, 142 pp. 1972
ISBN 3–540–05735–8

Vol. 12: T. Ando, M. Yamasaki, K. Suzuki
Protamines
Isolation, Characterization, Structure and Function
24 figs. 17 tables. IX, 114 pp. 1973
ISBN 3–540–06221–1

Vol. 13: P. Jollès, A. Paraf
Chemical and Biological Basis of Adjuvants
24 figs. 41 tables. VIII, 153 pp. 1973
ISBN 3–540–06308–0

Vol. 14: **Micromethods in Molecular Biology**
Edited by V. Neuhoff
275 figs (2 in color). 23 tables. XV, 428 pp. 1973
ISBN 3–540–06319–6

Vol. 15: M. Weissbluth
Hemoglobin
Cooperativity and Electronic Properties
50 figs. VIII, 175 pp. 1974
ISBN 3–540–06582–2

Vol. 16: S. Shulman
Tissue Specifity and Autoimmunity
32 figs. XI, 196 pp. 1974
ISBN 3–540–06563–6

Vol. 17: Y.A. Vinnikov
Sensory Reception
Cytology, Molecular Mechanisms and Evolution
124 figs. (173 separate ill.). IX, 392 pp. 1974
ISBN 3–540–06674–8

Vol. 18: H. Kersten, W. Kersten
Inhibitors of Nucleic Acid Synthesis
Biophysical and Biochemical Aspects
73 figs. IX, 184 pp. 1974
ISBN 3–540–06825–2

Vol. 19: M.B. Mathews
Connective Tissue
Macromolecular Structure and Evolution
31 figs. XII, 318 pp. 1975
ISBN 3–540–07068–0

Vol. 20: M.A. Lauffer
Entropy-Driven Processes in Biology
Polymerization of Tobacco
Mosaic Virus Protein and Similar Reactions
90 figs. X, 264 pp. 1975
ISBN 3–540–06933–X

Vol. 21: R.C. Burns, R.W.F. Hardy
Nitrogen Fixation in Bacteria and Higher Plants
27 figs. X, 189 pp. 1975
ISBN 3–540–07192–X

Vol. 22: H.J. Fromm
Initial Rate Enzyme Kinetics
88 figs. 19 tables. X, 321 pp. 1975
ISBN 3–540–07375–2

Prices are subject to change without notice

Springer-Verlag Berlin Heidelberg New York

Results and Problems in Cell Differentiation

A Series of Topical Volumes in Developmental Biology

Editors: W. Beermann; W. Gehring; J.B. Gurdon, F.C. Kafatos, J. Reinert

Vol. 1: **The Stability of the Differentiated State**
Editor: H. Ursprung
56 figs., XI, 144 pp. 1968.
ISBN 3-540-04315-2

Vol. 2: **Origin and Continuity of Cell Organelles**
Editors: J. Reinert; H. Ursprung
With contributions by numerous experts.
135 figs., XIII, 342 pp. 1971.
ISBN 3-540-05239-9

Vol. 3: **Nucleic Acid Hybridization in the Study of Cell Differentiation**
Editor: H. Ursprung
29 figs., XI, 76 pp. 1972.
ISBN 3-540-05742-0

Vo. 4: **Developmental Studies on Giant Chromosomes**
Editor: W. Beermann
110 figs., XV, 227 pp. 1972.
ISBN 3-540-05748-X

Vol. 5: **The Biology of Imaginal Disks**
Editors: H. Ursprung; R. Nöthiger
56 figs., XVII, 172 pp. 1972.
ISBN 3-540-05785-4

Vol. 6: W.J. Dickinson; D.T. Sullivan
Gene-Enzyme Systems in Drosophila
32 figs., XI, 163 pp. 1975.
ISBN 3-540-06977-1

Vol. 7: **Cell Cycle and Cell Differentiation**
J. Reinert; H. Holtzer
92 figs., XI, 331 pp. 1975.
ISBN 3-540-07069-9

Prices are subject to change without notice

Springer-Verlag Berlin Heidelberg New York